Operator Theory
Advances and Applications
Vol. 94

Editor
I. Gohberg

Series in Banach Spaces

Conditional and Unconditional Convergence

Mikhail I. Kadets
Vladimir M. Kadets

Translated from the Russian by Andrei Iacob

Birkhäuser Verlag
Basel · Boston · Berlin

Authors

M.I. Kadets and
V.M. Kadets
Pr. Pravdy 5, apt. 26
Kharkov 310022
Ukraine

1991 Mathematics Subject Classification 46B15, 46B20, 46G10, 42C20, 52A20, 10E05, 60B12

A CIP catalogue record for this book is available from the Library of Congress, Washington D.C., USA

Deutsche Bibliothek Cataloging-in-Publication Data
Kadec, Michail I.:
Series in Banach spaces : conditional and unconditional convergence / Mikhail I. Kadets ; Vladimir M. Kadets. Transl. from the Russ. by Andrei Iacob. - Basel ; Boston ; Berlin : Birkhäuser, 1997
(Operator theory ; Vol. 94)
ISBN 3-7643-5401-1 (Basel...)
ISBN 0-8176-5401-1 (Boston)
NE: Kadec, Vladimir M.:; GT

Printed on acid-free paper produced from chlorine-free pulp. TCF ∞
Cover design: Heinz Hiltbrunner, Basel
Printed in Germany
ISBN 3-7643-5401-1
ISBN 0-8176-5401-1

9 8 7 6 5 4 3 2 1

CONTENTS

INTRODUCTION

Series of scalars, vectors, or functions are among the fundamental objects of mathematical analysis. When the arrangement of the terms is fixed, investigating a series amounts to investigating the sequence of its partial sums. In this case the theory of series is a part of the theory of sequences, which deals with their convergence, asymptotic behavior, etc. The specific character of the theory of series manifests itself when one considers rearrangements (permutations) of the terms of a series, which brings combinatorial considerations into the problems studied.

The phenomenon that a numerical series can change its sum when the order of its terms is changed is one of the most impressive facts encountered in a university analysis course. The present book is devoted precisely to this aspect of the theory of series whose terms are elements of Banach (as well as other topological linear) spaces.

The exposition focuses on two complementary problems. The first is to characterize those series in a given space that remain convergent (and have the same sum) for any rearrangement of their terms; such series are usually called unconditionally convergent. The second problem is, when a series converges only for certain rearrangements of its terms (in other words, converges conditionally), to describe its sum range, i.e., the set of sums of all its convergent rearrangements.

Recall that for series of real numbers both problems are settled by a well-known theorem of Riemann (1867): a series converges unconditionally if and only if it converges absolutely; if the series converges conditionally, then its sum range is the whole real line.

Difficulties arise when one considers more general series, for instance, series in a finite-dimensional space, or even series of complex numbers. To be more precise, the answer to the first problem is readily obtained, as before: in any finite-dimensional vector space the commutativity law (invariance under rearrangements) holds only for absolutely convergent series. By contrast, the solution to the second problem—the description of the set of elements to which a series can converge for a suitable rearrangement of its terms—requires considerably more efforts. For series of complex numbers such a description was given by P. Lévy in 1905 [50], while for case of series in finite-dimensional spaces it was provided by E. Steinitz [87].

The study of rearrangements of series in infinite-dimensional spaces was initiated by W. Orlicz [63,64] and subsequently developed by many mathematicians in different countries, and continues to be an active field of research. However, until now this theme found almost no exposition in textbooks and monographs.

Several years ago we undertook the task of partially filling this gap by writing a textbook *Rearrangements of Series in Banach Spaces* [43], intended for students at Tartu University. However, the beautiful results of D. V. Pecherskii, S. A. Chobanyan, W. Banaszczyk, and others, which appeared after that textbook was published, changed in essential manner the state of the art of the theory of series, have led to the development of new approaches, and in some instances even to a new point of view on the subject. For this reason we have decided to address again, but on a more modern level, the same topics, the result being the present book.

One of our goals was to bring the topics under consideration and the theory of Banach spaces to the attention of young mathematicians. This explains the large number of exercises, among which are elementary as well as very difficult ones, and the fact that numerous open problems are pointed out. With the same goal in mind we included an Appendix titled *The limit set of the Riemann integral sums of a vector-valued function*, which, although does not belong directly to the main subject of the book, is closely related to it through the methods used in proofs and contains readily formulated open problems. We hope that our efforts will prove fruitful for beginner mathematicians as well for specialists in functional analysis and related areas of mathematics.

We deal here mostly with general series, but beyond the topics covered in our exposition there are many other situations where the idea of considering rearrangements of series has interesting applications: basic expansions (the theory of unconditional bases [12], [53], [83], [101]), orthogonal systems of functions, Fourier analysis, etc. As a sample we wish to mention here the following brilliant theorem of Sz. Gy. Révśz [100]: for every 2π-periodic continuous function f it is possible to rearrange its Fourier series in such a way that some subsequence of the sequence of partial sums of the rearranged series will converge to f uniformly.

For the understanding of the material presented in the book familiarity with standard courses of mathematical analysis and linear algebra and also with the elementary notions and results of the theory of Banach spaces is assumed (the content of the books [45], [55] or [78] is more than sufficient).

The junior of the authors has used the book to teach a special course at Khar'kov University. Discussions with the participants in the course were extremely useful in preparing the final version of the manuscript.

Our work was supported in part by the AMS fSU Aid Grants and, in the final stage, by ISSEP Grants SPU 061025 and APU 06140.

We wish to express our gratitude to W. Banaszczyk, I. S. Belov, W. A. Beyer, E. A. Gorin, V. S. Grinberg, and especially S. A. Chobanyan for useful discussions, to I. Halperin and T. Ando for kindly sending us supplementary bibliography on rearrangement of series, and to D. L. Dun, the spouse of the senior and the mother of the junior of the authors, who took upon herself the worries of everyday life—without the freedom she thus provided we would have never decided to undertake the work whose results are offered now to the reader.

NOTATIONS

In the book "Lemma 3.1.2" means "Lemma 2 in Section 1 of Chapter 3." Theorems, definitions, and exercises are numbered in a similar fashion, each type with its independent numbers, i.e., in the text one can encounter both, say, Definition 2.1.1 and Theorem 2.1.1.

The proof of any assertion starts with the word "Proof" and ends with the symbol $\square$. If $\square$ appears immediately after the formulation of an assertion, this means that the assertion is given without proof or that the proof follows from the preceding arguments.

Spaces and operators are as a rule denoted by upper case Italic letters, whereas elements of a space and linear functionals are denoted by lower case Italic letters. If not otherwise stipulated, "number" means "real number", and linear spaces are considered over the field $\mathbf{R}$ of real numbers.

The space $\mathbf{R}^n$ of rows of the form $x = (x_1, x_2, \ldots, x_n)$, $x_i \in \mathbf{R}$, is called here the n-dimensional coordinate space. The vectors $e_1 = (1, 0, 0, \ldots, 0)$, $e_2 = (0, 1, 0, \ldots, 0)$, $\ldots$, $e_n = (0, 0, \ldots, 0, 1)$ form the *standard basis* in $\mathbf{R}^n$. In infinite-dimensional spaces of sequences the unit (norm-one) vectors of the standard basis are also denoted by e_n.

Whenever it is clear what norm we speak about, we use the simple notation $\|\cdot\|$. When several norms in the same space are considered, we use notations like $\|\cdot\|_p$, $\|\cdot\|_X$, etc.

In the following list "—" stands for "denotes."

X^* — the dual (conjugate space) of the space X.

X^{**} — the second dual of the space X.

$\langle x, f\rangle$ — the value of the functional $f \in X^*$ on the element $x \in X$, or the inner product of two elements of Euclidean space.

$\operatorname{Lin} A$ — the linear span of the set A.

$\overline{\operatorname{Lin}} A$ — the closed linear span of the set A.

$B(X)$ — the unit ball of the normed space X, i.e., the set of elements whose norm is smaller than or equal to 1.

$S(X)$ — the unit sphere of the normed space X.

$\dim E$ — the dimension of the linear space E.

$l_p^{(n)}$, where $1 \leq p < \infty$ — the n-dimensional coordinate space, equipped with the norm

$$\|x\|_p = \left(\sum_{i=1}^{n} |x_i|^p\right)^{1/p},$$

where x_i are the coordinates of the vector x; $l_2^{(n)}$ will be called the Euclidean space.

$l_\infty^{(n)}$ — the n-dimensional coordinate space, equipped with the norm

$$\|x\|_\infty = \max_{1 \leq i \leq n} |x_i|.$$

$C[0,1]$ — the space of continuous functions on the segment $[0,1]$, with the norm $\|f\| = \max_{t \in [0,1]} |f(t)|$.

(Ω, Σ, μ) — a measure space, i.e., a set Ω with a σ-algebra Σ of measurable susbsets and a measure μ.

$L_p = L_p(\Omega, \Sigma, \mu)$, where $1 \leq p < \infty$ — the space of measurable functions $f \colon \Omega \to \mathbf{R}$ for which

$$\|f\| = \left(\int_\Omega |f|^p \, d\mu\right)^{1/p} < \infty.$$

$L_\infty = L_\infty(\Omega, \Sigma, \mu)$ — the space of bounded measurable functions $f \colon \Omega \to \mathbf{R}$, equipped with the norm

$$\|f\| = \operatorname*{ess\,sup}_{t \in \Omega} |f(t)|.$$

(Recall that when one deals with spaces of measurable functions one identifies functions that differ on a set of measure zero.) Here

$$\operatorname*{ess\,sup}_{t \in \Omega} f(t) = \inf\{\sup_{t \in U} f(t) \colon U \in \Sigma,\ \mu(\Omega \setminus U) = 0\}.$$

c_0 — the space of all numerical sequences $x = (x_1, x_2, \dots)$ that converge to 0, equipped with the norm $\|x\| = \max |x_i|$.

l_p, where $1 \le p < \infty$ — the space of all numerical sequences for which $\sum_{i=1}^{\infty} |x_i|^p < \infty$, equipped with the norm

$$\|x\| = \left(\sum_{i=1}^{\infty} |x_i|^p \right)^{1/p} .$$

l_∞ — the space of all bounded numerical sequences, equipped with the norm $\|x\| = \sup |x_i|$.

$\left(\sum_{n=1}^{\infty} X_n\right)_p$ — the l_p-sum of the spaces X_n, i.e., the space of all sequences $x = (x_1, x_2, \dots, x_n, \dots)$, $x_i \in X_i$, for which $\sum_{n=1}^{\infty} \|x_n\|^p < \infty$, equipped with the norm $\|x\| = \left(\sum_{n=1}^{\infty} \|x_n\|^p\right)^{1/p}$.

$\mathrm{SR}(\sum_{k=1}^{\infty} x_k)$ — the sum range of the series $\sum_{k=1}^{\infty} x_k$ (Definition 2.1.1).

$d(X, Y)$ — the Banach-Mazur distance between the spaces X and Y (Definition 5.1.1).

$X \xrightarrow{\mathrm{f}} Y$ — the space X is finitely representable in the space Y (Definition 5.1.2).

$X \overset{\mathrm{f}}{\Rightarrow} Y$ — the space Y is finitely saturated by the space X (Definition 6.3.4).

$\nabla f(x)$ — the Fréchet derivative of the function f at the point x (Definition 6.1.2).

CHAPTER 1
BACKGROUND MATERIAL

In this short chapter we collected facts that are sufficiently well-known, the consideration of which may be regarded as an introduction to the main text.

§1. Numerical Series. Riemann's Theorem

Since the basic properties of numerical series are well known from the calculus course, we will give here only the most needed definitions and facts.

DEFINITION 1.1.1. A series $\sum_{k=1}^{\infty} x_k$, $x_k \in \mathbf{R}$, is said to be *absolutely convergent* if the series $\sum_{k=1}^{\infty} |x_k|$ converges.

DEFINITION 1.1.2. A series $\sum_{k=1}^{\infty} x_k$, $x_k \in \mathbf{R}$, is said to be *conditionally convergent* if it converges, but the series $\sum_{k=1}^{\infty} |x_k|$ diverges.

DEFINITION 1.1.3. A *permutation* of a set A is a bijective mapping of A onto itself.

THEOREM 1.1.1. *Suppose that all x_k are nonnegative numbers and the series $\sum_{k=1}^{\infty} x_k$ converges to a number s. Then for any permutations π of the set of natural numbers the series $\sum_{k=1}^{\infty} x_{\pi(k)}$ converges and its sum is equal to s.*

PROOF. Let us denote by s_n the partial sum of the rearranged series: $s_n = \sum_{k=1}^{n} x_{\pi(k)}$. Since the terms are nonnegative, it is readily seen that $s_1 \le s_2 \le s_3 \le \dots\ s_n \le \dots$ and $\sup_k s_k \le s$. Hence, the sequence s_n converges to some number $\overline{s} \le s$, i.e., $\sum_{k=1}^{\infty} x_{\pi(k)} = \overline{s} \le s$. Switching the roles of the series $\sum_{k=1}^{\infty} x_k$ and $\sum_{k=1}^{\infty} x_{\pi(k)}$ and arguing in the same way as above, we obtain the opposite inequality $\sum_{k=1}^{\infty} x_k = s \le \overline{s} = \sum_{k=1}^{\infty} x_{\pi(k)}$. Therefore, $s = \overline{s}$. □

THEOREM 1.1.2. *Suppose that the series $\sum_{k=1}^{\infty} x_k$ is absolutely convergent. Then $\sum_{k=1}^{\infty} x_k$ converges; moreover, for any permutation π of the natural numbers the rearranged series $\sum_{k=1}^{\infty} x_{\pi(k)}$ converges and $\sum_{k=1}^{\infty} x_k = \sum_{k=1}^{\infty} x_{\pi(k)}$.*

PROOF. For any real number x set $x^+ = x$ if $x > 0$ and $x^+ = 0$ if $x \le 0$, and set $x^- = x^+ - x$. Consider $\sum_{k=1}^{\infty} x_k^+$ and $\sum_{k=1}^{\infty} x_k^-$. These are series of nonnegative numbers; they are convergent, being dominated by the series $\sum_{k=1}^{\infty} |x_k|$. Denote $\sum_{k=1}^{\infty} x_k^+ = s^+$ and $\sum_{k=1}^{\infty} x_k^- = s^-$. Then, by Theorem 1.1.1, for any permutation π the series $\sum_{k=1}^{\infty} x_{\pi(k)}^+$ and $\sum_{k=1}^{\infty} x_{\pi(k)}^-$ converge; moreover, $\sum_{k=1}^{\infty} x_{\pi(k)}^+ = s^+$ and

$\sum_{k=1}^{\infty} x_{\pi(k)}^{-} = s^{-}$. Consequently, the series $\sum_{k=1}^{\infty} x_{\pi(k)} = \sum_{k=1}^{\infty} \left(x_{\pi(k)}^{+} - x_{\pi(k)}^{-}\right)$ converges, and $\sum_{k=1}^{\infty} x_{\pi(k)} = s^{+} - s^{-}$. Clearly, the initial series $\sum_{k=1}^{\infty} x_k$ also converges, and its sum is equal to $s^{+} - s^{-}$. □

THEOREM 1.1.3 (RIEMANN'S THEOREM). *Let $\sum_{k=1}^{\infty} x_k$ be a conditionally convergent series of real numbers. Then:*

1) *for any $s \in \mathbf{R}$ one can find a permutation π such that $\sum_{k=1}^{\infty} x_{\pi(k)} = s$;*

2) *one can find a permutation σ such that $\sum_{k=1}^{\infty} x_{\sigma(k)} = \infty$.*

PROOF. Let us partition the set $\mathbf{N}$ of natural numbers into two sets, $A = \{a_1, a_2, \dots\}$ and $B = \{b_1, b_2, \dots\}$, such that $x_{a_k} \geq 0$, $x_{b_k} < 0$ for all $k \in \mathbf{N}$. But if $\sum_{k=1}^{\infty} x_{a_k} < \infty$ and $\sum_{k=1}^{\infty} x_{b_k} > -\infty$, then the series $\sum_{k=1}^{\infty} |x_k|$ converges. But if $\sum_{k=1}^{\infty} x_{a_k} = \infty$ and $\sum_{k=1}^{\infty} x_{b_k} > -\infty$, or $\sum_{k=1}^{\infty} x_{a_k} < \infty$ and $\sum_{k=1}^{\infty} x_{b_k} = -\infty$, then the series $\sum_{k=1}^{\infty} x_k$ diverges. Since, by the hypothesis of the theorem, the series $\sum_{k=1}^{\infty} x_k$ converges, but not absolutely, we see that $\sum_{k=1}^{\infty} x_{a_k} = \infty$ and $\sum_{k=1}^{\infty} x_{b_k} = -\infty$. Now let us prove assertions 1) and 2).

1) Let $s \geq 0$. We construct the needed permutation π as follows: $\pi(1) = a_1$, $\pi(2) = a_2, \dots, \pi(j) = a_j$ and so on, up to the number $j = j_1$ for which the partial sum $s_j = \sum_{k=1}^{j} x_{\pi(k)}$ of the rearranged series becomes larger than s: $s_k \leq s$ for $0 < k \leq j_1 - 1$, and $s_{j_1} > s$. Then we start to add negative terms: $\pi(j_1+1) = b_1$, $\pi(j_1+2) = b_2$, ... up to the moment when the partial sum $s_{j_1+j_2}$ becomes for the first times smaller than s. Then we again add positive terms, then again negative ones, and so on. It is readily seen that the partial sums s_j constructed in this manner tend to s. In the case $s < 0$ the construction is almost identical.

2) This time the permutation σ is constructed as follows: $\sigma(1) = a_1$, $\sigma(2) = a_2, \dots$, up to j_1 for which s_{j_1} becomes larger than $1 - x_{b_1}$, and we put $\sigma(j_1) = b_1$. Then we again add positive terms for as long as s_{j_2} is not larger than $2 - x_{b_2}$, and we put $\sigma(j_2+1) = b_2$, and we continue in this way ad infinitum. Then $s_j > k$ for all $j > j_k$, which implies that $s_j \to \infty$. A similar procedure yields a permutation for which the permuted series diverges to $-\infty$. □

EXERCISE 1.1.1. Clarify whether Riemann's theorem is valid for series of complex numbers.

The results of this section, and among them the main one — Riemann's theorem —, give information on how a series of numbers behaves with respect to different permutations of its terms. At the same times there exists another, relatively poorly studied circle of problems, dealing with how one and the same permutation acts on different series. The exercises given below concern precisely this kind of problems.

DEFINITION 1.1.4. A permutation $\pi\colon \mathbf{N} \to \mathbf{N}$ is said to be *convergence-modifying* if there exists a convergent numerical series $\sum_{k=1}^{\infty} x_k$ for which the series $\sum_{k=1}^{\infty} x_{\pi(k)}$ diverges.

EXERCISE 1.1.2. Characterize the convergence-modifying permutations in terms of the "degree to which they mix" the index set.

DEFINITION 1.1.5. A permutation $\pi\colon \mathbf{N} \to \mathbf{N}$ is said to be *sum-modifying* if there exists a convergent numerical series $\sum_{k=1}^{\infty} x_k$ for which the series $\sum_{k=1}^{\infty} x_{\pi(k)}$ converges, but $\sum_{k=1}^{\infty} x_k \neq \sum_{k=1}^{\infty} x_{\pi(k)}$.

EXERCISE 1.1.3. Characterize the sum-modifying permutations.

EXERCISE 1.1.4. Show that every sum-modifying permutation is convergence-modifying.

EXERCISE 1.1.5. Let π be a permutation such that $\pi(3k) = 2k$, $\pi(3k+j) = 4k - 2j + 1$ for $k \in \mathbf{N}$ and $j = 1, 2$. Further, let σ be a permutation such that $\sigma(5k) = 2k$, $\sigma(5k + j) = 8k + 2j - 1$ for $k \in \mathbf{N}$ and $j = 1, 2, 3, 4$. Show that if the series $\sum_{k=1}^{\infty} x_k$ and $\sum_{k=1}^{\infty} x_{\pi(k)}$ converge and $\sum_{k=1}^{\infty} x_k = 0$, then the series $\sum_{k=1}^{\infty} x_{\sigma(k)}$ also converges, and $\sum_{k=1}^{\infty} x_{\sigma(k)} = 2\sum_{k=1}^{\infty} x_{\pi(k)}$.

§2. Main Definitions. Elementary Properties of Vector Series

In this section we included those definitions and properties of series of elements in Banach spaces which do not differ in essential manner from the corresponding definitions and properties of ordinary numerical series.

DEFINITION 1.2.1. A *series of elements of a Banach space* X is an expression having the form of the sum of an infinite number of terms that belong to X:

$$x_1 + x_2 + \cdots + x_n + \dots . \tag{1}$$

To simplify notation of series one uses the symbol $\sum_{k=1}^{\infty} x_k$. One has to remember that the expression (1) is not a sum in the usual sense of the word, because in a Banach space addition is defined only for a finite number of terms.

DEFINITION 1.2.2. The *n-th partial sum of the series* $\sum_{k=1}^{\infty} x_k$ is the sum $s_n = \sum_{k=1}^{n} x_k$ of its first n ($n < \infty$) terms.

DEFINITION 1.2.3. A series is said to be *convergent* if the sequence of its partial sums converges in the norm of the space X. The limit of that sequence is called the *sum of the series*: $s = \lim_{n\to\infty} s_n$. Whenever we write $s = \sum_{k=1}^{\infty} x_k$ we mean that the series $\sum_{k=1}^{\infty} x_k$ converges and its sum is equal to s.

DEFINITION 1.2.4. The *remainder* of the convergent series $\sum_{k=1}^{\infty} x_k$ is the element $r_n = s - s_n$. In other words, the remainder r_n is the sum of the series $x_{n+1} + x_{n+2} + \dots$.

When n grows the remainder tends to zero. Sometimes the term "remainder" is used not for the sum of the series $\sum_{k=n+1}^{\infty} x_k$, but for the series $x_{n+1} + x_{n+2} + \dots$ itself.

DEFINITION 1.2.5. A *segment* of a series is a sum of a finite number of consecutive terms: $\sum_{k=m+1}^{n} x_k = s_n - s_m$. Sometimes by a "segment" one means the set $\{x_k\}_{k=m+1}^{n}$ itself.

THEOREM 1.2.1 (THE CAUCHY CONVERGENCE CRITERION). *The series $\sum_{k=1}^{\infty} x_k$ converges if and only if the sequence (more precisely — the net) of its segments converges to zero:*

$$\lim_{m,n\to\infty} \left\| \sum_{k=m+1}^{n} x_k \right\| = 0.$$

PROOF. Suppose the series converges: $s_n \to s$ as $n \to \infty$. Then

$$\left\| \sum_{k=m+1}^{n} x_k \right\| = \|s_n - s_m\| \le \|s_n - s\| + \|s_m - s\| \to 0,$$

i.e., Cauchy's condition is satisfied.

Conversely, suppose Cauchy's condition holds: $\lim_{m,n\to\infty} \|s_n - s_m\| = 0$. This means that the partial sums form a Cauchy (fundamental) sequence. Since the space X is complete, any fundamental sequence in X converges; hence, the sequence of partial sums s_n converges. □

DEFINITION 1.2.6. The series $\sum_{k=1}^{\infty} x_k$ is said to be *absolutely convergent* if $\sum_{k=1}^{\infty} \|x_k\| < \infty$.

THEOREM 1.2.2. *If the series $\sum_{k=n+1}^{\infty} x_k$ converges absolutely, then it converges.*

PROOF. Since $\sum_{k=1}^{\infty} \|x_k\| < \infty$, the Cauchy criterion yields

$$\lim_{m,n\to\infty} \sum_{k=m+1}^{n} \|x_k\| = 0.$$

By the triangle inequality,

$$\left\| \sum_{k=m+1}^{n} x_k \right\| \le \sum_{k=m+1}^{n} \|x_k\|,$$

whence $\lim_{m,n\to\infty} \| \sum_{k=m+1}^{n} x_k \| = 0$. By Cauchy's criterion, this means that the series $\sum_{k=1}^{\infty} x_k$ converges. □

It is not difficult to show (this is left as an exercise for the reader) that if the series $\sum_{k=1}^{\infty} x_k$ and $\sum_{k=1}^{\infty} y_k$ converge, then the series $\sum_{k=1}^{\infty}(ax_k + by_k)$ also converges; and if $a_k \in \mathbf{R}_+$, $k = 1, 2, \ldots$, $\sum_{k=1}^{\infty} a_k < \infty$, and $\|x_k\| \le a_k$ for all k, then the series $\sum_{k=1}^{\infty} x_k$ converges. Further, if $T\colon X \to Y$ is a continuous linear operator, then for any convergent series $\sum_{k=1}^{\infty} x_k$ of elements of X the series $\sum_{k=1}^{\infty} Tx_k$ also converges. Many other classical convergence tests for numerical series can also be readily generalized to the case of series in a Banach space.

EXERCISE 1.2.1. Show that if Theorem 1.2.1 or Theorem 1.2.2 holds in the normed space X, then X is complete (i.e., a Banach space).

§3. Preliminary Material on Rearrangements of Series of Elements of a Banach Space

From this moment on we shall address problems in which the specific features of the underlying space, its geometric and linear-topological properties, can play an important role.

DEFINITION 1.3.1. A series $\sum_{k=1}^{\infty} x_k$ is said to be *unconditionally convergent* if it converges for any rearrangement of its terms.

A numerical series is unconditionally convergent if and only if it is absolutely convergent. From Theorem 1.2.2 it readily follows that in one direction this connection remains valid in the general case as well: absolute convergence in a Banach space implies unconditional convergence. The converse is false, as the next example shows.

EXAMPLE 1.3.1. Let $X = l_2$ and $x_k = (0, 0, \ldots, 0, k^{-1}, 0, \ldots)$, where the nonzero coordinate is the kth. Then the series $\sum_{k=1}^{\infty} x_k$ converges to the element $s = (1, 2^{-1}, 3^{-1}, \ldots, n^{-1}, \ldots)$ for any rearrangement of its terms, but does not converge absolutely, since $\sum_{k=1}^{\infty} \|x_k\| = \sum_{k=1}^{\infty} k^{-1} = \infty$.

A little down the road we shall see (the Dvoretsky-Rogers theorem in Chapter 4) that in every infinite-dimensional Banach space one can construct an unconditionally convergent series which is not absolutely convergent.

DEFINITION 1.3.2. A series $\sum_{k=1}^{\infty} x_k$ is said to be *conditionally convergent* if it converges, but not unconditionally, i.e., among its rearrangements there are divergent ones.

Notice that for numerical series we gave a different definition of the notion of conditional convergence (Definition 1.1.2); however, since for numerical series absolute and unconditional convergence are equivalent, those two definitions are equivalent.

THEOREM 1.3.1. *If the series $\sum_{k=1}^{\infty} x_k$ in the Banach space X is unconditionally convergent, then all its rearrangements have the same sum.*

PROOF. We argue by reductio ad absurdum. Let $\sum_{k=1}^{\infty} x_k = s$ and suppose that for some permutation π the sum s' of the rearranged series is not equal to s. Pick a functional $f \in X^*$ such that $f(s) \neq f(s')$. Then the numerical series $\sum_{k=1}^{\infty} f(x_k)$ will not converge absolutely, since the permutation π changes its sum. Hence, by Riemann's theorem, there exists a permutation σ for which the series $\sum_{k=1}^{\infty} f(x_{\sigma(k)})$ diverges. Then the series $\sum_{k=1}^{\infty} x_{\sigma(k)}$ will also diverge, which contradicts the unconditional convergence of the series $\sum_{k=1}^{\infty} x_k$. □

The study of unconditionally convergent series based directly on Definition 1.3.1 encounters some difficulties. For this reason we shall presently give another, more convenient definition, which is equivalent to the basic definition of unconditional convergence.

DEFINITION 1.3.3. A series $\sum_{k=1}^{\infty} x_k$ of elements of a Banach space is said to be *perfectly convergent* if the series $\sum_{k=1}^{\infty} \alpha_k x_k$ converges for any choice of the coefficients $\alpha_i = \pm 1$.

THEOREM 1.3.2. *For series $\sum_{k=1}^{\infty} x_k$ in a Banach space X the following conditions are equivalent:*

1) *the series converges unconditionally;*

2) *all series of the form $x_{n_1} + x_{n_2} + x_{n_3} + \dots$ (subseries of the series $\sum_{k=1}^{\infty} x_k$), where $n_1 < n_2 < n_3 < \dots$ converge;*

3) *the series converges perfectly.*

PROOF. 1) $\Rightarrow$ 2). Suppose 2) does not hold, i.e., there exists a sequence $n_1 < n_2 < n_3 < \dots$ for which the series $\sum_{i=1}^{\infty} x_{n_i}$ diverges. Then, by Cauchy's criterion, there exist $\varepsilon > 0$ and numbers $m_1 < r_1 < m_2 < r_2 < \dots$ such that $\left\| \sum_{i=m_k}^{r_k} x_{n_i} \right\| \geq \varepsilon$. Denote the terms of the series $\sum_{k=1}^{\infty} x_k$ that do not appear in any of the segments $\{x_{n_i}\}_{i=m_k}^{r_k}$ by $y_1, y_2, y_3, \dots$. Now construct a rearrangement of the series $\sum_{k=1}^{\infty} x_k$ as follows: write first the segment $\{x_{n_i}\}_{i=m_1}^{r_1}$, then the term y_1, then the segment $\{x_{n_i}\}_{i=m_2}^{r_2}$, then the term y_2, and so on. By Cauchy's criterion, the rearranged series will diverge, which contradicts the unconditional convergence of the initial series.

2) $\Rightarrow$ 1). Suppose 1) does not hold, and let $\sum_{k=1}^{\infty} x_{\pi(k)}$ be a divergent rearrangement of our series. By Cauchy's criterion, there exist disjoint segments Δ_i, $i \in \mathbf{N}$, of the rearranged series for which the infimum of the norms is larger than $\varepsilon > 0$. Permute the terms in each Δ_i again, rearranging them in the order of increase of their indices (i.e., in the original order). Denote the index of the first [resp. last] term in Δ_i by m_i [resp. r_i]: $\Delta_i \subset \{x_k\}_{k=m_i}^{r_i}$. Passing, if necessary, to a subsequence, one can easily choose Δ_i so that $r_1 < m_2 < r_2 < m_3 < r_3 < \dots$. Then, writing successively all terms from Δ_1, then from Δ_2, from Δ_3, and so on, one obtains a subseries of the series $\sum_{k=1}^{\infty} x_k$ that diverges, contrary to condition 2).

2) $\Rightarrow$ 3). Let $\{\alpha_i\}_1^{\infty}$ be an arbitrary sequence of ± 1. Partition the natural numbers as $\mathbf{N} = A \cup B$, where $A = \{n_1, n_2, \dots\}$ is the set of indices n for which $\alpha_n = 1$ and $B = \{m_1, m_2, \dots\}$ is the set of indices n for which $\alpha_n = -1$. Then by hypothesis both series $\sum_{k=1}^{\infty} x_{n_k}$ and $\sum_{j=1}^{\infty} x_{m_j}$ converge, and so the series $\sum_{i=1}^{\infty} \alpha_i x_i = \sum_{k=1}^{\infty} x_{n_k} - \sum_{j=1}^{\infty} x_{m_j}$ will also converge.

3) $\Rightarrow$ 2). Let $n_1 < n_2 < n_3 < \dots$ be arbitrary indices. Put $\{n_1, n_2, n_3, \dots\} = A$ and $\mathbf{N} \setminus A = B$. Consider two sequences of coefficients: $\alpha_i = 1$ for all $i \in \mathbf{N}$ and $\beta_i = 1$ for $i \in A$, $\beta_i = -1$ for $i \in B$. By hypothesis, the series $\sum_{i=1}^{\infty} \alpha_i x_i$ and $\sum_{i=1}^{\infty} \beta_i x_i$ converge, and consequently so does the series $x_{n_1} + x_{n_2} + x_{n_3} + \dots = \sum_{i=1}^{\infty} \frac{1}{2}(\alpha_i x_i + \beta_i x_i)$. □

Thus, we have established the equivalence of unconditional and perfect convergence. From now on we shall make no distinction between these two notions.

THEOREM 1.3.5. *Let X be an n-dimensional normed space. Then in X any unconditionally convergent series is absolutely convergent.*

PROOF. Since in a finite-dimensional space all norms are equivalent to one another, one can assume that $X = l_1^{(n)}$. Denote by f_i, $i = 1, \dots, n$, the coordinates functionals, which assign to each vector its i-th component. Let $\sum_{k=1}^{\infty} x_k$ be an unconditionally convergent series in X. Then the numerical series $\sum_{k=1}^{\infty} f_i(x_k)$ are also unconditionally convergent, and hence $\sum_{k=1}^{\infty} |f_i(x_k)| < \infty$ for $i = 1, \dots, n$. Therefore,

$$\sum_{k=1}^{\infty} \|x_k\| = \sum_{k=1}^{\infty} \left(\sum_{i=1}^{n} |f_i(x_k)| \right) < \infty$$

and the series $\sum_{k=1}^{\infty} x_k$ converges absolutely. □

The next assertion shows that unconditionally convergent series $\sum_{i=1}^{\infty} x_i$ not only have the property that all series $\sum_{i=1}^{\infty} x_i$ with $\alpha_i = \pm 1$ converge, but this convergence is in a certain sense uniform with respect to the coefficients α_i.

LEMMA 1.3.1. *Suppose the series $\sum_{i=1}^{\infty} x_i$ in a Banach space X converges unconditionally. Then for any $\varepsilon > 0$ there exists a number $N = N_\varepsilon \in \mathbf{N}$ such that*

$$\sup \left\{ \left\| \sum_{k=N}^{N+m} \alpha_i x_i \right\| : m \in \mathbf{N},\ \alpha_i = \pm 1 \right\} < \varepsilon.$$

PROOF. Suppose that the assertion of the lemma is false, i.e., there exist $\varepsilon > 0$ and natural numbers $m_1 < r_1 < m_2 < r_2 < \dots$ such that, for any $j \in \mathbf{N}$,

$$\max \left\{ \left\| \sum_{k=m_j}^{r_j} \alpha_i x_i \right\| : \alpha_i = \pm 1 \right\} \geq \varepsilon. \tag{1}$$

Define a sequence $\beta_i = \pm 1$, $i \in \mathbf{N}$, in such a way that

$$\left\| \sum_{k=m_j}^{r_j} \beta_i x_i \right\| = \max \left\{ \left\| \sum_{k=m_j}^{r_j} \alpha_i x_i \right\| : \alpha_i = \pm 1 \right\},$$

i.e., for i such that $m_j \leq i \leq r_j$ take for the β_i those α_i for which the maximum in (1) is attained, and for the remaining values of i take, say, $\beta_i = 1$. By hypothesis, the series $\sum_{i=1}^{\infty} \beta_i x_i$ must converge (the series $\sum_{i=1}^{\infty} x_i$ is unconditionally convergent); but inequality (1) shows that the constructed series diverges. The contradiction we have reached proves the lemma. □

THEOREM 1.3.4 (GEL'FAND'S THEOREM). *Let X be a Banach space and $\sum_{k=1}^{\infty} x_i$ be an unconditionally convergent series in X. Then the collection of the sums $s(\alpha) = \sum_{i=1}^{\infty} \alpha_i x_i$, where $\alpha = \{\alpha_i\}_{i=1}^{\infty}$ runs through all sequences of ± 1, forms a compact set in X.*

PROOF. Equip the set of sequences $\alpha = \{\alpha_i\}_{i=1}^{\infty}$ in question with the topology of coordinate-wise convergence. The resulting topological space is a compact space homeomorphic to the Tikhonov product $K = \{-1,1\}^{\mathbf{N}}$. By Lemma 1.3.1, the mapping $\alpha \mapsto s(\alpha) = \sum_{i=1}^{\infty} \alpha_i x_i$ from K to X is continuous in this topology. To complete the proof it remains to use the theorem asserting that the image of a compact space under a continuous mapping is compact. □

EXERCISE 1.3.1. Determine for which of the spaces l_p one can construct examples of unconditionally, but not absolutely convergent series, following the pattern of Example 1.3.1.

EXERCISE 1.3.2. Construct an unconditionally, but not absolutely convergent series in the space $L_1[0,1]$. Consider other classical Banach spaces.

EXERCISE 1.3.3. Show that if, in the definition of perfect convergence, instead of the convergence of the series $\sum_{i=1}^{\infty} \alpha_i x_i$ for all collections of coefficients $\alpha_i = \pm 1$, one requires the convergence of the series $\sum_{i=1}^{\infty} \theta_i x_i$ for all collections of coefficients $\theta_k \in T$, where $T \subset \mathbf{R}$ is a bounded set that contains at least two points, then one obtains an equivalent notion of convergence.

EXERCISE 1.3.4. Investigate for which classes of linear topological spaces one can successfully generalize the definitions given in the last two sections.

CHAPTER 2

SERIES IN A FINITE-DIMENSIONAL SPACE

We shall now turn to the exposition of the main material, starting with the "simplest" case where the space is finite-dimensional. Already here there arise rather nontrivial problems, which continue to be the subject of fruitful investigations. Analogies with the material studied in the present chapter will play an important role in the sequel, in the examination of features that are specific for the infinite-dimensional case.

§1. Steinitz's Theorem on the the Sum Range of a Series

DEFINITION 2.1.1. Let $\sum_{k=1}^{\infty} x_k$ be a series in a Banach space X. We say that a point x *belongs to the sum range of the series* if there exists a permutation π for which the rearranged series $\sum_{k=1}^{\infty} x_{\pi(k)}$ converges to x. The set of all such points, called the *sum range of the series* $\sum_{k=1}^{\infty} x_k$, will be denoted by $\mathrm{SR}(\sum_{k=1}^{\infty} x_k)$.

By Riemann's theorem, if $X = \mathbf{R}$, then $\mathrm{SR}(\sum_{k=1}^{\infty} x_k) = X$ for any conditionally convergent series in X. If the dimension of the space X is larger than one, then there exist conditionally convergent series whose sum range does not coincide with the whole space. For example, if all terms of a series are collinear to some vector e, then for any rearrangement the sum of the series will also be collinear to e. In this example the sum range of the series is the line passing through 0 and the point e.

Let X be a finite-dimensional space. For any given subspace Y of X the reader will find no difficulty in constructing a series $\sum_{k=1}^{\infty} x_k$ in X for which $\mathrm{SR}(\sum_{k=1}^{\infty} x_k) = Y$. Further, observing that the relation

$$\mathrm{SR}(x_0 + \sum_{k=1}^{\infty} x_k) = x_0 + \mathrm{SR}(\sum_{k=1}^{\infty} x_k)$$

holds for any $x_0 \in X$, one concludes that in a finite-dimensional space the sum range of a series may be any translated (affine) subspace. As it turns out, this exhausts all possible situations, i.e., if $\sum_{k=1}^{\infty} x_k$ is a conditionally convergent series in a finite-dimensional space, then its sum range is an affine subspace of dimension larger than or equal to one. In the two-dimensional case this fact was first established by P. Lévy in 1905 [50]; a proof for arbitrary finite-dimensional spaces was given by E. Steinitz [87]. In this section we shall prove the theorem formulated above, following mainly Steinitz's scheme.

For the proof we need several auxiliary assertions. The first of these can be formulated as follows: the number of faces that meet at any given vertex of a polyhedron is not smaller than the dimension of the polyhedron.

LEMMA 2.1.1. *Let K be a polyhedron in $\mathbf{R}^n$ given by a system of linear equalities and inequalities:*

$$\begin{cases} f_i(x) = a_i, & i = 1, 2, \ldots, p, \\ g_j(x) \leq b_j, & j = 1, 2, \ldots, q, \end{cases}$$

where f_i and g_j are linear functionals. Let x_0 be a vertex (extreme point) of K and $A = \{j \colon g_j(x_0) = b_j\}$. Then the number of elements in A is not smaller than $n - p$.

PROOF. Suppose the contrary holds. Then the system of linear equations

$$\begin{cases} f_i(x) = 0, & i = 1, 2, \ldots, p, \\ g_j(x) = 0, & j \in A, \end{cases}$$

has less equations than unknowns (the coordinates of the vector x), and hence has a nonntrivial solution x_1. For sufficiently small $\varepsilon > 0$ the points $x_0 \pm \varepsilon x_1$ will belong to K, which contradicts the fact that x_0 is a vertex of the polyhedron. □

LEMMA 2.1.2 (ROUNDING-OFF-COEFFICIENTS LEMMA). *Let $\{x_i\}_{i=1}^n$ be a finite subset of an m-dimensional normed space, $\{\lambda_i\}_{i=1}^n$ be a set of scalar coefficients, $0 \leq \lambda_i \leq 1$, and $x = \sum_{i=1}^n \lambda_i x_i$. Then there exists a set of coefficients $\{\theta_i\}_{i=1}^n$, each θ_i equal to 0 or 1 (a set of rounded off coefficients) such that*

$$\left\| x - \sum_{i=1}^{n} \theta_i x_i \right\| \leq \frac{m}{2} \cdot \max_i \|x_i\|. \tag{1}$$

PROOF. If $n \leq m$, then it suffices to take $\theta_i = 0$ whenever $\lambda_i \leq 1/2$ and $\theta_i = 1$ whenever $\lambda_i > 1/2$. Now consider the case $n > m$. Let us introduce the auxiliary space of coefficients $\mathbf{R}^n$ and the polyhedron K in $\mathbf{R}^n$ given by the system of inequalities $0 \leq t_i \leq 1$, $i = 1, 2, \ldots, n$, and equalities $x = \sum_{i=1}^n t_i x_i$, where $(t_1, t_2, \ldots, t_n)$ are the coordinates of a vector in K. Since the polyhedron K is nonempty and bounded, there exists a vertex $T = (\bar{t}_1, \bar{t}_2, \ldots, \bar{t}_n) \in K$. Notice that the vector equality $x = \sum_{i=1}^n t_i x_i$ is a system of m scalar equations. Hence, by Lemma 2.1.1, among the coordinates of the point T there are $n - m$ that are equal to 0 or 1. Now define the numbers θ_i as follows: if $\bar{t}_i = 0$ or $\bar{t}_i = 1$, then $\theta_i = \bar{t}_i$; if $0 < \bar{t}_i \leq 1/2$, then $\theta_i = 0$; finally, if $1/2 < \bar{t}_i < 1$, then $\theta_i = 1$. We have

$$\left\| \sum_{i=1}^{n} \theta_i x_i - x \right\| = \left\| \sum_{i=1}^{n} \theta_i x_i - \sum_{i=1}^{n} \bar{t}_i x_i \right\| \leq \left(\sum_{i=1}^{n} |\theta_i - \bar{t}_i| \right) \cdot \max_i \|x_i\|.$$

Since $n-m$ of the numbers $|\theta_i - \bar{t}_i|$ are equal to zero, and the remaining ones are no larger than $1/2$, we conclude that

$$\left\| \sum_{i=1}^{n} \theta_i x_i - x \right\| \leq \frac{m}{2} \cdot \max_i \|x_i\|. \qquad \square$$

EXERCISE 2.1.1. Show that if $X = l_2^{(m)}$, then the constant $m/2$ in the right-hand side of inequality (1) can be replaced by $\sqrt{m}/2$.

LEMMA 2.1.3 (REARRANGEMENT LEMMA). *Suppose that in the m-dimensional normed space X there is given a finite set $\{x_i\}_{i=1}^n$ of vectors, whose sum is denoted by x. Then one can rearrange the elements of this set in such a way that for any natural $k \leq n$ the following inequality holds:*

$$\left\| \sum_{i=1}^{k} x_{\pi(i)} - \frac{k-m}{n} x \right\| \leq m \cdot \max_i \|x_i\|, \tag{2}$$

where π is the corresponding permutation of the index set $\{1, 2, \ldots, n\}$.

PROOF. Since inequality (2) is homogeneous, i.e., is preserved when all the vectors x_i are multiplied by the same constant, we may assume that the norms of the elements are bounded by 1: $\max_i \|x_i\| = 1$. It also clear that we only need to prove inequality (2) for $n > m$, since for $n \leq m$ it holds for any permutation. Let us construct by induction a chain of sets $\{1, 2, \ldots, n\} = A_n \supset A_{n-1} \supset \ \ldots \ A_m$ and numbers λ_k^i $(k = m, m+1, \ldots, n;\ i \in A_k)$, with the following properties:

$$\operatorname{card} A_k = k; \quad 0 \leq \lambda_k^i \leq 1, \quad \sum_{i \in A_k} \lambda_k^i = k - m; \quad \sum_{i \in A_k} \lambda_k^i x_i = \frac{k-m}{n} x. \tag{3}$$

The set A_n is already given. Indeed, for $\lambda_n^i = (n-m)/n$ properties (3) with $k = n$ hold. The induction is carried out by decreasing k. Suppose the set A_{k+1} and the set of coefficients $\{\lambda_{k+1}^i\}_{i \in A_{k+1}}$ are already constructed. Consider the set K of collections of numbers $\{\mu_i,\ i \in A_{k+1}\}$ that satisfy the conditions

$$0 \leq \mu_i \leq 1; \qquad \sum_{i \in A_{k+1}} \mu_i = k - m; \qquad \sum_{i \in A_{k+1}} \mu_i x_i = \frac{k-m}{n} x. \tag{4}$$

Then K is nonempty (one can take $\mu_i = \frac{k-m}{k-m+1} \lambda_{k+1}^i$) and is a convex polyhedron in the space $\mathbf{R}^{k+1}$ of vectors $\mu = \{\mu_i \colon i \in A_{k+1}\}$. It is readily seen that the conditions of Lemma 2.1.1 are satisfied with $p = m+1$ (the last vector equality in (4) is equivalent to m scalar equalities) and $q = 2(k+1)$.

The polyhedron K is a bounded set, since all μ_i belong to the segment $[0, 1]$. It follows that K has vertices. Let $\overline{\mu} = \{\overline{\mu}_i, i \in A_{k+1}\}$ be one of these vertices.

By Lemma 2.1.1, the set A of those i for which μ_i is equal to 0 or 1 has at least $(k+1)-(m+1)=k-m$ elements.

We claim that at least one of the numbers $\overline{\mu}_i$ is equal to zero. Indeed, if $\overline{\mu}_i = 1$ for all $i \in A$, then by the first two conditions in (4), $\operatorname{card} A = k-m$, and the remining numbers μ_i $(i \in A_{k+1} \setminus A)$ are equal to 0. If $\overline{\mu}_i = 1$ does not hold for all $i \in A$, then again among the numbers $\overline{\mu}_i$ there are some equal to 0.

Let j be an index such that $\overline{\mu}_j = 0$. Set $A_k = A_{k+1} \setminus \{j\}$ and $\lambda_k^i = \overline{\mu}_i$ $(i \in A_k)$. It is readily verified that conditions (3) are satisfied. Hence, our construction is complete.

The sought-for permutation π is defined as follows: for $i = m+1, m+2, \dots, n$ set $\pi(i)$ equal to the index j that was eliminated from the set A_i in the inductive construction of the set A_{i-1}, and for the remaining values of i define $\pi(i)$ in arbitrary manner.

Let us check that the permutation π constructed above satisfies inequality (2). For $k \le m$ this is obvious. For $k > m$, by conditions (3), we have

$$\left\| \sum_{i=1}^{k} x_{\pi(i)} - \frac{k-m}{n} \cdot x \right\| = \left\| \sum_{i \in A_k} x_i - \sum_{i \in A_k} \lambda_k^i x_i \right\|$$

$$= \left\| \sum_{i \in A_k} (1-\lambda_k^i) x_i \right\| \le \sum_{i \in A_k} (1-\lambda_k^i) = k-(k-m) = m. \qquad \square$$

The proof of Lemma 2.1.3 given above belongs to V. S. Grinberg and S. V. Sevast'yanov [27].

REMARK 2.1.1. If we remove the second term from the left-hand side of inequality (2), we obtain the following inequality, which is more convenient for our purposes:

$$\left\| \sum_{i=1}^{k} x_{\pi(i)} \right\| \le m \cdot \max_i \|x_i\| + (m+1) \left\| \sum_{i=1}^{k} x_i \right\|. \tag{5}$$

REMARK 2.1.2. The last lemma is often formulated as follows: Let X be a finite-dimensional space, and let $\{x_k\}_{k=1}^n$ be a set of vectors such that $\sum_{k=1}^n x_k = 0$. Then there exists a permutation π of the first n natural numbers such that

$$\max_{j \le n} \left\| \sum_{k=1}^{j} x_{\pi(k)} \right\| \le K \cdot \max_i \|x_i\|, \tag{6}$$

where K depends only on the space X.

In this formulation Lemma 2.1.3 is known as *Steinitz's lemma*. The *Steinitz constant* of the space X is defined as the infimum $K(X)$ of the constants K that can be taken in inequality (6). From Lemma 2.1.3 it follows that $K(X) \le \dim X$.

It is probable that by using the specific structure of a space X one could better estimate the quantity $K(X)$. For instance, it is not known what is the Steinitz constant of a Euclidean space; one would guess that $K(l_2^{(n)}) \leq \sqrt{n}$. It even not known what is the exact value of the Steinitz constant of a three-dimensional Euclidean space (see [1], [2] for the best known values).

DEFINITION 2.1.2. Let X be a Banach space and let $\sum_{k=1}^{\infty} x_k$ be a conditionally convergent series in X. A linear functional $f \in X^*$ is called a *convergence functional* for this series if $\sum_{k=1}^{\infty} |f(x_k)| < \infty$. The set of all convergence functionals of a series will be denoted by Γ. Also, we will denote by $\Gamma^{\perp} \subset X$ the annihilator of the set of convergence functionals:

$$\Gamma^{\perp} = \{x \in X \colon f(x) = 0 \text{ for all } f \in \Gamma\}.$$

Clearly, Γ and $\Gamma^{\perp}$ are linear subspaces of X^* and X, respectively. In the infinite-dimensional case Γ is not necessarily closed. Next let us introduce the following two sets:

$$\mathcal{P}(\{x_k\}_{k=1}^{\infty}) = \{x_{i_1} + x_{i_2} + \cdots + x_{i_p} \colon i_1 < i_2 < \cdots < i_p;\ p \in \mathbf{N}\},$$

$$\mathcal{Q}(\{x_k\}_{k=1}^{\infty}) = \left\{ \sum_{i=1}^{N} \lambda_i x_i \colon 0 \leq \lambda_i \leq 1,\ N = 1, 2, \ldots \right\}.$$

The elements of $\mathcal{P}(\{x_k\}_{k=1}^{\infty})$ will be referred to as *partial sums*. $\mathcal{P}(\{x_k\}_{k=1}^{\infty}) \subset \mathcal{Q}(\{x_k\}_{k=1}^{\infty})$; also, the set $\mathcal{Q}(\{x_k\}_{k=1}^{\infty})$ is convex. We let $\overline{\mathcal{Q}}$ denote the closure of $\mathcal{Q}(\{x_k\}_{k=1}^{\infty})$.

LEMMA 2.1.4. *Let X be an arbitary Banach space, and let $\sum_{k=1}^{\infty} x_k$ be a conditionally convergent series in X. Then for any $x \in \overline{\mathcal{Q}}$ the set $x + \Gamma^{\perp}$ is contained in $\overline{\mathcal{Q}}$.*

PROOF. Pick an arbitrary functional $f \in X^* \setminus \Gamma$. Clearly, for such an f the numerical series $\sum_{k=1}^{\infty} f(x_k)$ converges and its sum equals $f(\sum_{k=1}^{\infty} x_k)$, but the convergence is not absolute: $\sum_{k=1}^{\infty} |f(x_k)| = \infty$. It follows that among the partial sums of the series $\sum_{k=1}^{\infty} f(x_k)$ there are arbitrarily large ones "in both directions," i.e.,

$$\sup\{f(y) \colon y \in \mathcal{P}(\{x_k\}_{k=1}^{\infty})\} = \infty,$$
$$\inf\{f(y) \colon y \in \mathcal{P}(\{x_k\}_{k=1}^{\infty})\} = -\infty.$$

Since $\mathcal{Q}(\{x_k\}_{k=1}^{\infty}) \supset \mathcal{P}(\{x_k\}_{k=1}^{\infty})$, we see that

$$\sup\{f(y) \colon y \in \mathcal{Q}(\{x_k\}_{k=1}^{\infty})\} = \infty. \tag{7}$$

Now suppose the conclusion of the lemma is false, i.e., there exist $x \in \overline{\mathcal{Q}}$ and $z \in \Gamma^{\perp}$ such that $x + z \notin \overline{\mathcal{Q}}$. Then, by the Hahn-Banach theorem, one can separate $x + z$ from the set $\overline{\mathcal{Q}}$ by some linear functional $f \in X^*$:

$$\sup\{f(y) \colon y \in \overline{\mathcal{Q}}\} < f(x + z).$$

If $f \in \Gamma$, then the last condition cannot be satisfied, since then $f(z) = 0$, and consequently $f(x) = f(x+z)$. If now $f \notin \Gamma$, then again the last condition cannot be satisfied, since in this case (7) yields $\sup\{f(y)\colon y \in \overline{Q}\} = \infty$. We have reached a contradiction. □

Lemma 2.1.4 is taken from V. P. Fonf's paper [23].

REMARK 2.1.3. Let S' be an arbitrary finite subset of terms of the series $\sum_{k=1}^{\infty} x_k$, and let y be an arbitrary element of the space X. Then the conclusion of the preceding lemma remains valid if we replace $\overline{Q}(\{x_k\}_{k=1}^{\infty})$ by $y + \overline{Q}(\{x_k\}_{k=1}^{\infty} \setminus S')$.

We proved the last lemma under more general assumptions than we actually need now, with the future aim of establishing an infinite-dimensional generalization of Steinitz's theorem. In the present section Lemma 2.1.4 will actually be used only for series in an m-dimensional space.

THEOREM 2.1.1 (STEINITZ'S THEOREM). *Let $\sum_{k=1}^{\infty} x_k$ be a convergent series in an m-dimensional space E, and let $\sum_{k=1}^{\infty} x_k = s$. Then the sum range of the series is the affine subspace $s + \Gamma^{\perp}$, where $\Gamma^{\perp}$ is the annihilator of the set Γ of convergence functionals:* $\mathrm{SR}(\sum_{k=1}^{\infty} x_k) = s + \Gamma^{\perp}$.

PROOF. Let $f \in X^*$ be a convergence functional for $\sum_{k=1}^{\infty} x_k$. By definition, $\sum_{k=1}^{\infty} f(x_k)$ converges absolutely. Hence, for any permutation π, if the series $\sum_{k=1}^{\infty} x_{\pi(k)}$ converges, then

$$f\left(\sum_{k=1}^{\infty} x_{\pi(k)}\right) = \sum_{k=1}^{\infty} f(x_{\pi(k)}) = f(s).$$

Therefore, $\mathrm{SR}(\sum_{k=1}^{\infty} x_k) - s \subset \operatorname{Ker} f$. Since the last inclusion holds for all $f \in \Gamma$, we conclude that $\mathrm{SR}(\sum_{k=1}^{\infty} x_k) - s \subset \Gamma^{\perp}$, i.e., $\mathrm{SR}(\sum_{k=1}^{\infty} x_k) \subset s + \Gamma^{\perp}$.

The proof of the opposite inclusion $\mathrm{SR}(\sum_{k=1}^{\infty} x_k) \supset s + \Gamma^{\perp}$ is considerably more difficult and will be broken into two steps. First we will show that for any element $s' \in s + \Gamma^{\perp}$ one can find a permutation π_0 of the given series and a sequence of indices $n_1 < n_2 < \ldots$ such that

$$\lim_{j\to\infty} \left\| s' - \sum_{k=1}^{n_j} x_{\pi_0(k)} \right\| = 0,$$

i.e., that some subsequence of the sequence of partial sums of the rearranged series converges to s'. Only after that we will construct a "corrected" permutation π for which the series $\sum_{k=1}^{\infty} x_{\pi(k)}$ will indeed converge to s'.

The set $\overline{Q}$ contains s and, by Lemma 2.14, it also contains s' ($s' \in s + \Gamma^{\perp} \subset \overline{Q}$). Take a sequence of numbers $\varepsilon_n \searrow 0$. Approximate s' by an element $q_1 \in Q(\{x_k\}_{k=1}^{\infty})$: $\|s' - q_1\| = \|s' - \sum \lambda_i x_i\| < \varepsilon_1$. Next, applying the Rounding-off-Coefficients Lemma (Lemma 2.1.2), approximate q_1 by an element

$p_1 \in \mathcal{P}(\{x_k\}_{k=1}^{\infty}) : \|q_1 - \sum \theta_i x_i\| \le m \cdot \max_{k\ge 1} \|x_k\|$, where $m = \dim E$ and the coefficients θ_i are equal to 0 or 1. Consider the set of those elements x_k for which $\theta_k = 1$ and adjoin to it the element x_1, it x_1 is not already there. Denote the resulting set by S_1 and the sum of its elements by s_1. Then

$$\|s' - s_1\| \le \varepsilon_1 + m \cdot \max_{k\ge 1} \|x_k\| + \|x_1\|.$$

Now consider the set $s_1 + \overline{\mathcal{Q}}(\{x_k\}_{k=1}^{\infty} \setminus S_1)$. It contains s, and, by Remark 2.1.3, $s' \in s_1 + \overline{\mathcal{Q}}(\{x_k\}_{k=1}^{\infty} \setminus S_1)$. Further, approximate $s' - s_1$ by an element $q_2 \in \overline{\mathcal{Q}}(\{x_k\}_{k=1}^{\infty} \setminus S_1)$: $\|s' - s_1 - q_2\| = \|s' - s_1 - \sum \lambda_i x_i\| \le \varepsilon_2$. Now approximate q_2 by an element $p_2 \in \overline{\mathcal{P}}(\{x_k\}_{k=1}^{\infty} \setminus S_1)$: $\|q_2 - p_2\| = \|q_2 - \sum \theta_i x_i\| \le m \cdot \max_{k\ge 2} \|x_k\|$, where the θ_i's are equal to 0 or 1. Adjoin to S_1 all elements x_i from the last sum for which $\theta_i = 1$ and also the element x_2, if x_2 is not already in S_1 or among the newly adjoined elements. Denote the resulting set by S_2 and the sum of its elements by s_2. Then

$$\|s' - s_2\| \le \varepsilon_2 + m \cdot \max_{k\ge 2} \|x_k\| + \|x_2\|.$$

Continuing this construction indefinitely we obtain a sequence of finite sets

$$S_1 \subset S_2 \subset S_3 \subset \dots, \quad \bigcup_n S_n = \{x_k\}_{k=1}^{\infty},$$

such that the sums s_n of their elements satisfy the inequality

$$\|s' - s_n\| \le \varepsilon_n + m \cdot \max_{k\le n} \|x_k\| + \|x_n\|.$$

If we now write the elements of the sets $S_1,\ S_2 \setminus S_1,\ S_3 \setminus S_2,\ S_4 \setminus S_3, \dots$ one after another, then, by the estimates derived above, we obtain the desired rearrangement of the series $\sum_{k=1}^{\infty} x_k$.

Now let us proceed to the second step of our proof. We have produced a series (for convenience we will denote it again by $\sum_{k=1}^{\infty} x_k$) whose general term converges to zero (thanks to the convergence of the initial, unrearranged series), and for which a subsequence of its partial sums converges to s':

$$\lim_{j\to\infty} \left\| s' - \sum_{k=1}^{n_j} x_k \right\| = 0, \quad n_1 < n_2 < n_2 < n_3 < \dots .$$

From the last relation it follows that

$$\lim_{j\to\infty} \left\| \sum_{k=n_j+1}^{n_{j+1}} x_k \right\| = 0.$$

Apply to each of the sets $\{x_k\}_{k=n_j+1}^{n_{j+1}}$ the Rearrangement Lemma in the form given in Remark 2.1.1, and denote the resulting permutation of the set $\mathbf{N}$ by π. Then

$$\sum_{k=1}^{n_j} x_{\pi(k)} = \sum_{k=1}^{n_j} x_k, \quad j = 1, 2, \ldots,$$

and, for any $l \in (n_j, n_{j+1}]$,

$$\left\| \sum_{k=n_j+1}^{l} x_{\pi(k)} \right\| \leq m \cdot \max_{n_j < k \leq n_j+1} \|x_k\| + (m+1) \left\| \sum_{k=n_j+1}^{n_{j+1}} x_k \right\|.$$

Let r be an arbitrary natural number, $r > n_1$, and let j be such that $n_j < r \leq n_{j+1}$. Then

$$\left\| \sum_{k=1}^{r} x_{\pi(k)} - s' \right\| \leq \left\| \sum_{k=1}^{n_j} x_{\pi(k)} - s' \right\| + \left\| \sum_{k=n_j+1}^{r} x_{\pi(k)} \right\|$$

$$\leq \left\| \sum_{k=1}^{n_j} x_k - s' \right\| + m \cdot \max_{k > n_1} \|x_k\| + (m+1) \left\| \sum_{k=n_j+1}^{n_{j+1}} x_k \right\|.$$

Consequently,

$$\lim_{r \to \infty} \left\| \sum_{k=1}^{r} x_{\pi(k)} - s' \right\| = 0.$$

This means that the series $\sum_{k=i}^{\infty} x_{\pi(k)}$ converges and its sum is equal to s'. □

REMARK 2.1.4. From the last theorem it follows that the sum range of a conditionally convergent series in a finite-dimensional space cannot reduce to a single point. We leave the proof of this assertion to the reader.

In this section we have encountered a group of problems that could be regarded as belonging more to combinatorial geometry than to functional analysis, which, generally speaking, should come as no surprize. Indeed, here the main subject of our investigations is the permutation of vectors, i.e., a combinatorial notion. If one follows such an terminology, then the main object studied in this book is combinatorial geometry in an infinite-dimensional space.

As is sometimes the case, results of some or another abstract branch of mathematics turn out to be useful in applied problems. Thus, to us it came as a complete surprize the fact that the Rounding-off-Coefficients Lemma and the Rearrangement Lemma are of essential use in a number of optimal design problems (see [8], [82], and also [43, pp. 21–22]). From this point of view the estimation of the quantity $K(X)$ (see Remark 2.1.2) is not only an interesting and difficult problem, but also one of practical importance.

EXERCISE 2.1.2. Show that if a series $\sum_{k=1}^{\infty} x_k$ in a Banach space X converges, if $\sum_{k=1}^{\infty} x_k = 0$, and if for some permutation π one has that $\sum_{k=1}^{\infty} x_{\pi(k)} = x$, then one can find a permutation σ and a sequence of indices $n_1 < n_2 < n_3 < \dots$ such that $\lim_{j\to\infty} \sum_{k=1}^{n_j} x_{\sigma(k)} = 2x$.

EXERCISE 2.1.3. Empolying only the preceding exercise and the Rearrangement Lemma, show that in a finite-dimensional space the sum range of a series is an affine subspace.

EXERCISE 2.1.4. Characterize the sequences $\{x_k\}_{k=1}^{\infty}$ in $\mathbf{R}^n$ which can be ordered in such a way that their elements will form a convergent series. Solve first the one-dimensional version of the problem.

§2. The Dvoretzky-Hanani Theorem on Perfectly Divergent Series

In this section we shall consider the series that from the point of view of their properties are the farthest from the unconditionally convergent series.

DEFINITION 2.2.1. A series $\sum_{k=1}^{\infty}$ is said to be *perfectly divergent* if for any arrangement of signs the series $x_1 \pm x_2 \pm x_3 \pm \dots$ diverges.

It is clear that if the general term of a series does not tend to zero, then the series is perfectly divergent. The simple example of the series $\sum_{n=1}^{\infty} n^{-1/2} e_n$, where e_n is an orthonormal sequence in a Hilbert space, shows that failure of the general term to tend to zero is not the only reason for the perfect divergence of a series. Nevertheless, in a finite-dimensional space the perfectly divergent series are precisely the series whose general term does not tend to zero. Our next goal is to prove this result, obtained by A. Dvoretzky and C. Hanani in 1947 [18].

LEMMA 2.2.1. *Let $\{x_i\}_{i=1}^{n}$ be elements of a space X, $\dim X = m$. Then there exist coefficients $\alpha_i = \pm 1$, $i = 1, \dots, n$, such that*

$$\max_{j\le n} \left\| \sum_{i=1}^{j} \alpha_i x_i \right\| \le 2m \cdot \max_{1\le i\le n} \|x_i\|.$$

PROOF. (The proof given here belongs to V. S. Grinberg.) If the number n of elements does not exceed the dimension m of the space, then by the triangle inequality the above inequality holds for any choice of signs α_i. Now let $n > m$. Consider the polyhedron K_1 in $\mathbf{R}^{m+1}$ given by the system of equalities $\sum_{i=1}^{m+1} t_i x_i = 0$ and inequalities $-1 \le t_i \le 1$, $i = 1, \dots, m+1$, where $(t_i)_{i=1}^{m+1}$ denote the coordinates of a vector in K_1. This bounded polyhedron is not empty (indeed, $0 \in K_1$). Hence, by Lemma 2.1.1, there exists a collection $(t_i^1)_{i=1}^{m+1} \in K_1$ such that at least one of the coordinates, which we denote by t_{i_1}, is equal to 1 or -1. Now consider the polyhedrom K_2 in $\mathbf{R}^{m+2}$ given by the equalities $t_{i_1} = t_{i_1}^1$, $\sum_{i=1}^{m+2} t_i x_i = 0$, and the inequalities $1 \le t_i \le 1$, $i \in \{1, \dots, m+2\} \setminus \{i_1\}$. By Lemma 2.1.1, there exists a collection $(t_i^2)_{i=1}^{m+2} \in K_2$ for which at least one of the coordinates, which we denote by $t^2_{i_2}$, is equal to 1 or -1 $(i_2 \neq i_1)$. Moreover, $t^2_{i_1} = t^1_{i_1}$ by construction.

Repeating this process $n - m$ times, we obtain indices $i_1, i_2, \ldots, i_{n-m}$ and collections of coefficients $(t^k_i)^{m+k}_{i=1}$ such that $i_k \leq m + k$, $\sum^{m+k}_{i=1} t^k_i x_i = 0$, $|t^k_i| \leq 1$, $|t^k_{i_k}| = 1$, and $t^k_{i_j} = t^j_{i_j}$ for $k > j$. The sought-for collection of signs α_i can now be defined as follows: $\alpha_{i_k} = t^k_{i_k}$, and if i is different from all the numbers i_k, put $\alpha_i = 1$. Then for any $j \leq m$ we have

$$\left\| \sum^j_{i=1} \alpha_i x_i \right\| \leq \sum^j_{i=1} \|x_i\| \leq m \cdot \max_i \|x_i\|.$$

Now suppose $j > m$; set $j = m + l$. By construction, the sum $\sum^{m+l}_{i=1}(\alpha_i - t^l_i)x_i$ contains no more than m nonzero terms. Consequently, if we take into account that $\sum^{m+l}_{i=1} t^l_i x_i = 0$, we obtain

$$\left\| \sum^{m+l}_{i=1} \alpha_i x_i \right\| = \left\| \sum^{m+l}_{i=1} (\alpha_i - t^l_i) x_i \right\| \leq 2m \cdot \max_i \|x_i\|.$$

Hence, for any $j \leq n$,

$$\left\| \sum^j_{i=1} \alpha_i x_i \right\| \leq 2m \cdot \max_i \|x_i\|,$$

as claimed. □

REMARK. As in Steinitz's Lemma, here the best value of the constant that one can put instead $2m$ in the right-hand side of the inequality obtained above is not known. The case $X = l^{(m)}_2$ is of special interest.

THEOREM 2.2.1 (THE DVORETZKY-HANANI THEOREM). *If a series in a finite-dimensional normed space is perfectly divergent, then its general term does not tend to zero.*

PROOF. We shall proceed by reductio ad absurdum. Let X be an m dimensional space, and let $\sum^\infty_{n=1} x_n$ be a series in X such that $\lim_{n\to\infty} x_n = 0$. We need to exhibit an arangement of signs for which the series $\sum^\infty_{n=1} \pm x_n$ will converge. Denote $d_n = \sup_{i>n} \|x_i\|$. Clearly, $d_n \downarrow 0$. Consider a sequence of indices $0 = n_1 < n_2 < n_3 < \ldots$ such that $\sum^\infty_{k=1} d_{n_k} < \infty$. Using Lemma 2.21, for each of the segments $\{x_i\}^{n_{k+1}}_{i=n_k+1}$ one can choose signs t_i such that

$$\max_{n_k < j \leq n_{k+1}} \left\| \sum^j_{i=n_k+1} t_i x_i \right\| \leq 2m d_{n_k}.$$

Then, if $n_k < j_1 < j_2$, we have $\left\| \sum^{j_2}_{i=j_1} t_i x_i \right\| \leq 4m \sum^\infty_{j=k} d_j$. That is to say, for our choice of signs the series $\sum^\infty_{i=1} t_i x_i$ satisfies the Cauchy criterion, and hence converges. □

In any infinite-dimensional Banach space there are perfectly divergent series whose general term tends to zero. The reader will be able to prove this result on his own by solving the next exercise and using Dvoretzky's theorem on almost Euclidean subspaces (Theorem 6.2.1).

EXERCISE 2.2.1. For a Banach space X the following two assertions are equivalent:

1) for any perfectly divergent series in X the general term does not tend to zero;

2) there exists a constant $C > 0$ such that, for any finite set of elements $\{x_i\}_{i=1}^n \subset X$ there exists a set of coefficients $t_i = \pm 1$ (depending on the set $\{x_i\}_{i=1}^n$), for which

$$\max_{j \leq n} \left\| \sum_{i=1}^{j} t_i x_i \right\| \leq C \max_i \|x_i\|.$$

The following interesting problem remains open: Does the Dvoretzky-Hanani theorem extend to series in nuclear Fréchet spaces?

EXERCISE 2.2.2. Does the assertion of Exercise 2.2.1 hold true for noncomplete normed spaces?

§3. Pecherskii's Theorem

In 1988 D. V. Pecherskii [68] published the following result which, in particular, establishes a connection between the Dvoretzky-Hanani and the Steinitz theorems.

THEOREM 2.3.1. *Let X be an arbitrary Banach space, $\sum_{n=1}^{\infty} x_n$ be a conditionally convergent series in X, and $x_0 = \sum_{n=1}^{\infty} x_n$. Further, suppose that no rearrangement makes the series perfectly divergent. Then $\mathrm{RS}(\sum_{n=1}^{\infty} x_n)$ is the closed affine subspace $x_0 + \Gamma^{\perp}$, where $\Gamma^{\perp}$ is the anihilator in X of the set $\Gamma \subset X^*$ of convergence functionals (the terminology is taken from Definitions 2.1.1 and 2.1.2).*

This assertion provides the most general of the known sufficient conditions for linearity of the sum range of a series in an infinite-dimensional space. In the finite-dimensional case Theorem 2.3.1 is identical to Steinitz's theorem, because, by the Dvoretzky-Hanani theorem, the perfect divergence of a series is not affected by rearrangements of its terms. Pecherskii's theorem was later, although independently, also proved by S. A. Chobanyan. Since Chobanyan's proof is simpler than Pecherskii's proof, we shall follow here his scheme.

In the proof of Steinitz's theorem the finite-dimensionality of the space was used twice: in the Rearrangement Lemma 2.3.1 and in the Rounding-off-Coefficients Lemma 2.1.2. Analyzing that proof, one can derive the following assertion, which already holds in any Banach space.

LEMMA 2.3.1. *Suppose the series $\sum_{n=1}^{\infty} x_n$ in a Banach space X has the following two properties:*

(A) *for any $\varepsilon > 0$ there exist $N = N(\varepsilon)$ and $\delta > 0$ such that if $\{y_i\}_{i=1}^{n}$ is a finite set of terms of the series, $\{y_i\}_{i=1}^{n} \subset \{x_i\}_{i=N}^{\infty}$, $\|\sum_{i=1}^{n} y_i\| \leq \delta$, then one can find a permutation π of the first n natural numbers for which*

$$\max_{j \leq n} \left\| \sum_{i=1}^{j} y_{\pi(j)} \right\| \leq \varepsilon;$$

(B) *for any $\varepsilon > 0$ there exists a number $M = M(\varepsilon)$ such that if $\{y_i\}_{i=1}^{n}$ is a finite set of terms of the series, $\{y_i\}_{i=1}^{n} \subset \{x_i\}_{i=M}^{\infty}$, and if $0 \leq \lambda_i \leq 1$, $i = 1, \dots, n$, then one can find a set of coefficients $\{\theta_i\}_{i=1}^{n}$, $\theta_i \in \{0, 1\}$, for which*

$$\left\| \sum_{i=1}^{n} \lambda_i y_i - \sum_{i=1}^{n} \theta_i y_i \right\| \leq \varepsilon.$$

Then for any $x \in \mathrm{RS}(\sum_{n=1}^{\infty} x_n)$ it holds that $\mathrm{RS}(\sum_{n=1}^{\infty} x_n) = x + \Gamma^{\perp}$. □

The aim of the following lemmas is to show that under the assumptions of Theorem 2.3.1 the series $\sum_{n=1}^{\infty} x_n$ satisfies conditions (A) and (B) of Lemma 2.3.1.

LEMMA 2.3.2. *Suppose that there is no rearrangement that makes the series $\sum_{n=1}^{\infty} x_n$ perfectly convergent. Then for any $\varepsilon > 0$ there exists a number $N = N(\varepsilon)$ such that, for any finite collection, written in arbitrary order, of elements $\{y_1, y_2, \dots y_n\}$ of the set $\{x_N, x_{N+1}, x_{N+2}, \dots\}$ one can find a collection of signs $\alpha_i = \pm 1$ for which*

$$\max_{j \leq n} \left\| \sum_{i=1}^{j} \alpha_i y_i \right\| \leq \varepsilon.$$

PROOF. Suppose the assertion of the lemma is false. Then one can find an $\varepsilon > 0$ such that, for any $N \in \mathbf{N}$, the set $\{x_i\}_{i=1}^{\infty}$ contains a finite subset $\{y_i\}_{i=1}^{n}$ with the property that

$$\max_{j \leq n} \left\| \sum_{i=1}^{j} \alpha_i y_i \right\| > \varepsilon$$

for any choice of signs $\alpha_i = \pm 1$. Since such a subset $\{y_i\}_{i=1}^{n}$ exists in a tail of arbitrarily high order of the series, one can find numbers $1 = N_1 < N_2 < \dots$ and collections of terms of the series $\{y_i^k\}_{i=1}^{n_k} \subset \{x_i\}_{i=N_k}^{N_{k+1}-1}$, such that

$$\min_{\alpha_i = \pm 1} \max_{j \leq N_k} \left\| \sum_{i=1}^{j} \alpha_i y_i^k \right\| > \varepsilon. \tag{1}$$

Now write the terms of the series in the following order:

$$\{y_i^1\}_{i=1}^{n_1}, \quad \{x_i\}_{N_1}^{N_2-1} \setminus \{y_i^1\}_{i=1}^{n_1}, \quad \{y_i^2\}_{i=1}^{n_2}, \quad \{x_i\}_{N_2}^{N_3-1} \setminus \{y_i^2\}_{i=1}^{n_2}, \quad \dots,$$

where the elements of the sets $\{y_i^k\}_{i=1}^{n_k}$ are written in exactly the order for which inequality (1) holds, while the elements of the sets $\{x_i\}_{N_k}^{N_{k+1}-1} \setminus \{y_i^k\}_{i=1}^{k_k}$ are written in arbitrary order. Denote the resulting rearrangement of the series $\sum_{i=1}^{\infty} x_i$ by π. By inequality (1), the series $\sum_{i=1}^{\infty} \alpha_i x_{\pi(i)}$ diverges for any choice of signs α_i, which contradicts the hypothesis. □

LEMMA 2.3.3. *Let ε be an arbitrary positive number, and let $\{x_i\}_{i=1}^N$ be a set of elements of the normed space X, with the property that for any finite subset $\{y_i\}_{i=1}^m \subset \{x_i\}_{i=1}^N$ there exist signs $\alpha_i = \pm 1$ such that $\|\sum_{i=1}^m \alpha_i y_i\| \le \varepsilon$. Then for any set of coefficients $\{\lambda_i\}_1^N$, $0 \le \lambda_i \le 1$ there exist "rounded off" coefficients $\theta_i \in \{0,1\}$ for which*

$$\left\| \sum_{i=1}^N \lambda_i x_i - \sum_{i=1}^N \theta_i x_i \right\| \le \varepsilon.$$

PROOF. From considerations of continuity it is clear that it suffices to prove the lemma in the case when the coefficients λ_i are finite dyadic fractions:

$$\lambda_i = \lambda_{i,0} + \lambda_{i,1} \cdot \frac{1}{2} + \lambda_{i,2} \cdot \frac{1}{4} + \ldots + \lambda_{i,n} \cdot \frac{1}{2^n}, \tag{2}$$

where the $\lambda_{i,k}$ are equal to 0 or 1. The number of terms in the representation (2) will be called the *length* of the fraction. It is clear that if $\lambda_{i,0} = 1$, then because of the inequalities $0 \le \lambda_i \le 1$ all the remaining $\lambda_{i,k}$'s in (2) are equal to zero, and also that all λ_i may be regarded as dyadic fractions of the same length (adding, if needed, more zero terms).

Let us show, by induction on n, that if λ_i are fractions of length n, then one can choose $\theta_i \in \{0,1\}$ such that

$$\left\| \sum_{i=1}^N \lambda_i x_i - \sum_{i=1}^N \theta_i x_i \right\| \le \varepsilon \left(1 - \frac{1}{2^{n-1}}\right). \tag{3}$$

This will complete the proof of the lemma.

$n = 1$. In this case the λ_i themselves are 0 or 1, and we obtain zero in the left-hand side of inequality (3) if we take $\theta_i = \lambda_i$.

The $n \to n+1$ step. Let μ_i be fractions of length $n+1$, i.e., $\mu_i = \mu_{i,0} + \mu_{i,1} \cdot 2^{-1} + \ldots + \mu_{i,n} \cdot 2^{-n}$. Denote by A the set of all indices $i \le N$ for which $\mu_{i,n} = 1$. By hypothesis, there exist signs $\{\alpha_i\}_{i \in A}$ for which $\|\sum_{i \in A} \alpha_i x_i\| \le \varepsilon$. Introduce auxiliary numbers λ_i as follows: if $i \notin A$, put $\lambda_i = \mu_i$; if $i \in A$, put $\lambda_i = \mu_i + \alpha_i \cdot 2^{-n}$. Then

$$\left\| \sum_{i=1}^N \lambda_i x_i - \sum_{i=1}^N \mu_i x_i \right\| \le \frac{1}{2^n} \left\| \sum_{i \in A} \alpha_i x_i \right\| \le \varepsilon \cdot \frac{1}{2^n}.$$

On the other hand, the λ_i are dyadic fractions of length n, and by the inductive hypothesis there exist numbers $\theta_i \in \{0,1\}$ such that inequality (3) holds. This yields

$$\left\|\sum_{i=1}^N \mu_i x_i - \sum_{i=1}^N \theta_i x_i\right\| \leq \left\|\sum_{i=1}^N \lambda_i x_i - \sum_{i=1}^N \theta_i x_i\right\| + \left\|\sum_{i=1}^N \mu_i x_i - \sum_{i=1}^N \lambda_i x_i\right\|$$
$$\leq \varepsilon\left(1 - \frac{1}{2^{n-1}}\right) + \varepsilon \cdot \frac{1}{2^n} = \varepsilon\left(1 - \frac{1}{2^n}\right). \qquad \square$$

Combining lemmas 2.3.2 and 2.3.3, we conclude that under the assumptions of Theorem 2.3.1 condition (B) of Lemma 2.3.1 —an analogue of the Rounding-off-Coefficients Lemma—holds. It remains to establish property (A)—the analogue of the Rearrangement Lemma.

LEMMA 2.3.4 (CHOBANYAN'S LEMMA). *Let $\{x_i\}_{i=1}^n$ be elements of the space X, with $\sum_{i=1}^n x_i = 0$. Then there exists a permutation σ such that, for any choice of signs $\alpha_i = \pm 1$,*

$$\max_{k\leq n}\left\|\sum_{i=1}^k \alpha_i x_{\sigma(i)}\right\| \geq \max_{k\leq n}\left\|\sum_{i=1}^k x_{\sigma(i)}\right\|. \tag{4}$$

PROOF. Take σ to be the permutation for which

$$\min_\nu\left\{\max_{k\leq n}\left\|\sum_{i=1}^k x_{\nu(i)}\right\|\right\}$$

is attained. Let α_i be an arbitrary fixed collection of signs. We claim that (4) holds. Indeed, denote by A [resp. B] the set of indices i for which $\alpha_i = 1$ [resp. $\alpha_i = -1$], and put $A_k = A \cap \{1,\dots,k\}$, $B_k = B \cap \{1,\dots,k\}$, $B_k^c = \{1,\dots,n\} \setminus B_k$. Then

$$\max_{k\leq n}\left\|\sum_{i=1}^k \alpha_i x_{\sigma(i)}\right\| = \max_{k\leq n}\left\|\sum_{i\in A_k} x_{\sigma(i)} - \sum_{i\in B_k} x_{\sigma(i)}\right\|.$$

Consequently,

$$\max_{k\leq n}\left\|\sum_{i=1}^k \alpha_i x_{\sigma(i)}\right\| + \max_{k\leq n}\left\|\sum_{i=1}^n x_{\sigma(i)}\right\|$$
$$\geq 2\max\left\{\max_{k\leq n}\left\|\sum_{i\in A_k} x_{\sigma(i)}\right\|, \max_{k\leq n}\left\|\sum_{i\in B_k} x_{\sigma(i)}\right\|\right\}$$
$$= 2\max\left\{\max_{k\leq n}\left\|\sum_{i\in A_k} x_{\sigma(i)}\right\|, \max_{k\leq n}\left\|\sum_{i\in B_k^c} x_{\sigma(i)}\right\|\right\} \tag{5}$$

(to obtain the last equality we used the condition $\sum_{i=1}^{n} x_i = 0$). Since we have the inclusions $A_1 \subset A_2 \subset \cdots \subset A_n \subset B_n^c \subset B_{n-1}^c \subset \cdots \subset B_1^c$, of which n are equalities, there exists a permutation ν such that

$$\max_{k\le n}\left\|\sum_{i=1}^{k} x_{\nu(i)}\right\| = \max\left\{\max_{k\le n}\left\|\sum_{i\in A_k} x_{\sigma(i)}\right\|, \max_{k\le n}\left\|\sum_{i\in B_k^c} x_{\sigma(i)}\right\|\right\}. \tag{6}$$

By the definition of the permutation σ,

$$\max_{k\le n}\left\|\sum_{i=1}^{k} x_{\nu(i)}\right\| \ge \max_{k\le n}\left\|\sum_{i=1}^{n} x_{\sigma(i)}\right\|. \tag{7}$$

Combining (5), (6), and (7) we conclude that

$$\max_{k\le n}\left\|\sum_{i=1}^{k} \alpha_i x_{\sigma(i)}\right\| + \max_{k\le n}\left\|\sum_{i=1}^{n} x_{\sigma(i)}\right\| \ge 2\max_{k\le n}\left\|\sum_{i=1}^{n} x_{\sigma(i)}\right\|,$$

which in turn yields the needed inequality (4). □

LEMMA 2.3.5. *Let $\{x_i\}_{i=1}^n \subset X$, $\|\sum_{i=1}^n x_i\| \le \varepsilon$, and assume that for any permutation ν there exists a choice of signs $\alpha_i = \pm 1$ for which*

$$\max_{k\le n}\left\|\sum_{i=1}^{k} \alpha_i x_{\nu(i)}\right\| \le \varepsilon.$$

Then there exists a permutation σ such that $\max_{k\le n}\|\sum_{i=1}^{k} x_{\sigma(i)}\| \le 3\varepsilon$.

PROOF. Denote $x_{n+1} = -\sum_{i=1}^n x_i$. Then $\sum_{i=1}^{n+1} x_i = 0$, and we can apply the preceding lemma to the set $\{x_i\}_{i=1}^{n+1}$. This yields a permutation $\pi : \{1, \ldots, n+1\} \to \{1, \ldots, n+1\}$ such that

$$\max_{k\le n+1}\left\|\sum_{i=1}^{k} x_{\pi(i)}\right\| \le \min_{\alpha_i=\pm1}\max_{k\le n+1}\left\|\sum_{i=1}^{k} \alpha_i x_{\pi(i)}\right\|.$$

If we now discard the element x_{n+1}, we obtain a permutation σ of $\{1, \ldots, n\}$, for which

$$\max_{k\le n}\left\|\sum_{i=1}^{k} x_{\sigma(i)}\right\| \le \min_{\alpha_i=\pm1}\max_{k\le n}\left\|\sum_{i=1}^{k} \alpha_i x_{\sigma(i)}\right\| + 2\|x_{n+1}\| \le 3\varepsilon,$$

as claimed. □

Combining lemmas 2.3.2 and 2.3.5, we conclude that under the assumptions of Theorem 2.3.1 an analogue of the Rearrangement Lemma—assertion (A) of Lemma 2.3.1—holds true. Since property (B) has been established earlier, this completes the proof of Pecherskii's theorem.

EXERCISE 2.3.1. Give an example of a convergent series in a Hilbert space which after a rearrangement becomes perfectly divergent.

CHAPTER 3
CONDITIONAL CONVERGENCE IN AN INFINITE-DIMENSIONAL SPACE

Throughout this chapter we shall consider mainly the classical L_p spaces (with the exception of §3). The transition to a more general situation will be made later, after an exposition of the necessary supplementary material.

§1. Basic Counterexamples

Now that in the preceding chapter we gave a complete description of the sum range of a conditionally convergent series in a finite-dimensional space, it is only natural to attempt to extend those results to the infinite-dimensional setting. However, as is most frequently the case, here the extension to the infinite-dimensional object of the properties of its finite-dimensional analogue cannot be achieved in full. For this reason in what follows, in order to obtain analogues of Steinitz's theorem, we will have (as, for instance, in Pecherskii's theorem) to impose additional restrictions on series. The main goal of this section is to construct examples demonstrating that without additional restrictions it is not possible to extend Steinitz's theorem to series in an infinite-dimensional Banach space.

We have already remarked that if $\sum_{k=1}^{\infty} x_k$ is a conditionally convergent series in a finite-dimensional space, then $\mathrm{SR}(\sum_{k=1}^{\infty} x_k)$ is an infinite set; more precisely, it contains a whole line. The next examples shows that for a conditionally convergent series in an infinite-dimensional space the sum range may consist of a single point.

EXAMPLE 3.1.1. The construction will be carried out in the Hilbert space l_2. Let $\{e_k\}_{k=1}^{\infty}$ be the standard orthonormal basis in l_2. Consider the series

$$e_1 + \frac{1}{2}e_2 - \frac{1}{2}e_2 + \frac{1}{4}e_3 - \frac{1}{4}e_3 + \frac{1}{4}e_3 - \frac{1}{4}e_3$$
$$+\frac{1}{8}e_4 - \frac{1}{8}e_4 + \frac{1}{8}e_4 - \frac{1}{8}e_4 + \frac{1}{8}e_4 - \frac{1}{8}e_4 + \frac{1}{8}e_4 - \frac{1}{8}e_4 + \frac{1}{16}e_5 - \dots .$$

This series converges and its sum is equal to e_1. However, it does not converge unconditionally, since the series in which all the signs are replaced by + diverges (recall that, by Theorem 1.3.2, perfect and unconditional convergence coincide). On the other hand, the sum range of our series reduces to the point e_1: indeed, the projection of the series on any coordinate axis contains only a finite number of nonzero terms, and a finite sum does not change under rearrangements.

EXERCISE 3.1.1. Show that in any infinite-dimensional Banach space there exists a conditionally convergent series whose sum range consists of a single point.

EXERCISE 3.1.2. Show that the expansion of the function $\ln(1 + x)$ in a Maclaurin series is a conditionally convergent series in the space $C[0, 1]$ and that the sum range of this series is a single point of $C[0, 1]$.

EXERCISE 3.1.2. Solve Exercise 2.3.1 by a construction analogous to that of Example 3.1.1.

The main part of the assertion of Steinitz's theorem is the linearity of the set $\mathrm{SR}(\sum_{k=1}^{\infty} x_k)$, i.e., the fact that $\mathrm{SR}(\sum_{k=1}^{\infty} x_k)$ contains, together with any two distinct points, the line that connects them. Below we shall discuss an example which shows that in an infinite-dimensional Hilbert space there are series with a nonlinear sum range.

It is a known historical fact that the mathematicians of the Lvov school had the habit of meeting in the "Scottish Café" and discuss there mathematical problems. At that time the theory of Banach spaces has been just created, and during these table talks the Lvov mathematicians, under the leadership of Stefan Banach, laid the foundations of modern functional analysis. When mathematicians discuss some topic, they often need to write down formulas, draw figures, etc. During the discussions at the "Scottish Café" such notes were written in indelible ink right on a marble table. After that the waiters would wipe off the notes. Some results of Banach's school were perhaps lost only because they were not copied from table to paper. This difficulty was solved by Łucja, Banach's wife. She bought a thick hardbound booknote and gave it to the café's cashier. (According to another version, Banach himself bought the booknote.) From that moment on people started to write the problems proposed (and, if they were successfully solved, sketches of their solutions) in the booknote. Any mathematician visiting the cafe could leave a note in the book. For the solution of some problems their authors would set a prize—from a small cup of coffee to a goose. In this way a manuscript was created, named the "Scottish Book" [81]. After the 2nd World War this book was published in the USA by one of Banach's students, S. Ulam. Some problems in the "Scottish Book" were solved after its publication, while others remain open even today; part of the unsolved problems are passed on as "folklore" and many people are not aware of the fact that the original source is the "Scottish Book." The topics dealt with in the "Scottish Book" determined in many respects the directions of further development of the theory of Banach spaces.

In problem 106 of the "Scottish Book" Banach proposed to prove that for any series in a Banach space its sum range is a linear set. A prize was offered for the solution of this problem — a bottle of wine. The solution also appears in the Book; it presumably belongs to J. Marcinkiewicz (authorship established according to handwriting). As already mentioned, the answer provided was negative, namely, a counterexample in the space $L_2[0, 1]$.

Unfortunately, Soviet mathematicians were not aware of this result, and it was rediscovered by E. M. Nikishin, who constructed a more complicated example.

It has to be said, however, that Nikishin's example simultaneously solved a considerably more difficult problem posed by Banach and Orlicz, concerning the sum range of a series of functions in the sense of almost-everywhere convergence. We shall return to this problem in the last chapter. Now let us describe Marcinkiewicz's construction.

Denote by $\chi_{[a,b]}$ the function that takes the value 1 on the segment $[a, b]$ and the value 0 on its complement (the indicator or characteristic function of the segment $[a, b]$).

EXAMPLE 3.1.2. Consider the series with the terms

$$x_{i,k} = \chi_{[k\cdot 2^{-i},(k+1)\cdot 2^{-i}]}; \quad y_{i,k} = -x_{i,k},$$

where $0 \leq i < \infty$, $0 \leq k < 2^i$; these are elements of $L_2[0, 1]$. If we write its sum in the order

$$x_{0,0} + y_{0,0} + x_{1,0} + y_{1,0} + +x_{1,1} + y_{1,1} + x_{2,0} + \dots,$$

then the series will converge to zero. If we now change the order to

$$x_{0,0} + x_{1,0} + x_{1,1} + y_{0,0} + x_{2,0} + x_{2,1} + y_{1,0} + x_{2,2} + x_{2,3} + y_{1,1} + \dots,$$

then the series will converge to the function identically equal to 1 on $[0, 1]$. However, no rearrangement will make the series converge to the function identically equal to $1/2$ on $[0, 1]$, because any of the partial sums of the series is an integer-valued function.

The next example, which belongs to M. I. Ostrovskii [65], shows that the set $\mathrm{SR}(\sum_{k=1}^{\infty} x_k)$ is not necessarily closed.

EXAMPLE 3.1.3. The example lives in the Hilbert space $L_2([0, 1] \times [0, 1])$ of square-integrable functions on the square $0 \leq x \leq 1$, $0 \leq y \leq 1$. Let $\sum_{k=1}^{\infty} f_k(t)$ be the series constructed in the preceding example. Our new series is constructed as follows:

$$\sum_{k=1}^{\infty} g_k(x, y) = f_1(x) + \sqrt{2} f_1(y) + f_2(x) + \sqrt{2} f_2(y) + \dots .$$

It is not hard to show that a function $u(x, y) \equiv \text{const}$ belongs to $\mathrm{SR}(\sum_{k=1}^{\infty} g_k)$ if and only if $u(x, y) = n + \sqrt{2}m$, where n and m are integers. Hence, the set $\mathrm{SR}(\sum_{k=1}^{\infty} g_k)$ is not closed.

EXERCISE 3.1.4. Show that the sum range of the series constructed in Example 3.1.2 contains all integer-valued functions.

EXERCISE 3.1.5. Show that if X is a Banach space and Y is a separable subspace of X, then there exists a series $\sum_{k=1}^{\infty} x_k$ for which $\mathrm{SR}(\sum_{k=1}^{\infty} x_k) = Y$.

EXERCISE 3.1.6. Show that by means of an appropriate rearrangement the series of Example 3.1.2 can be made to converge weakly to any prescribed element of L_2.

EXERCISE 3.1.7. Determine for which function spaces on the segment $[0,1]$ does the series of functions given in Example 3.1.2 have a convergent rearrangement.

EXERCISE 3.1.8. In each of the spaces l_p and c_0 exhibit examples series with nonlinear sum range.

REMARK 3.1.1. There is no known example of a series in a Hilbert space whose sum range is a linear, but nonclosed subspace.

REMARK 3.1.2. As we will show in the last chapter, there are infinite-dimensional linear topological spaces in which Steinitz's theorem is valid. In the case of normed spaces the validity of Steinitz's theorem implies the finite-dimensionality of the space (Chapter 7, § 2).

EXERCISE 3.1.9. Show that Steinitz's theorem holds in the space of all numerical sequences with the topology of coordinate-wise convergence.

§2. A Series Whose Sum Range Consists of Two Points

In this section we shall give an example of a series in a space L_p (which consisting of the same functions for all values of p, $1 \le p < \infty$), whose sum range consists of two points — the functions identically equal to 0 and to 1. This example disproves, in particular, H. Hadwiger's conjecture that the sum range of any conditionally convergent series is the coset of some additive subgroup of the space under consideration.

The construction given here belongs to M. I. Kadets; its justification was first obtained independently, and about the same time, by P. A. Kornilov [48] and K. Wozniakowski (see [34]). It is interesting that similar constructions were proposed at least by two other mathematicians. A. Dvoretzky told us that many years ago he had such an example, but he, too (like M. I. Kadets), was not able to find a justification. P. Enflo constructed an example with a complete proof at about the same time [48] and [34] were written, but he did not publish it because I. Halperin informed him about the preprint containing the example presented below.

Denote by $Q = [0,1]^\omega$ the infinite dimensional cube, i.e., the product of a countable number of unit segments, equipped with the standard product probability measure. The example will be constructed in $L_p(Q)$, $1 \le p < \infty$.

Our series consists of functions of two kinds. The functions of the first kind are defined as follows:

$$f_m^n(t) = f_m^n(t_2, t_2, \dots) = \begin{cases} 1 & \text{if } \frac{m-1}{n} < t_n < \frac{m}{n} \\ 0 & \text{otherwise.} \end{cases} \tag{1}$$

for $n \in \mathbf{N}$, $m = 1, 2, \ldots, n$. For each $n \in \mathbf{N}$ set $F_n = \{f_m^n : 1 \leq m \leq n\}$. The functions of the second kind are defined as products:

$$g_{m,j}^n = -f_m^n \cdot f_j^{n+1} \quad \text{for } n \in \mathbf{N}, \quad 1 \leq m \leq n, \quad 1 \leq j \leq n+1. \tag{2}$$

For each $n \in \mathbf{N}$ set $G_n = \{g_{m,j}^n : 1 \leq m \leq n,\ 1 \leq j \leq n+1\}$. Clearly, $\|f_m^n\| = n^{-1/p}$, $\|g_{m,j}^n\| = (n(n+1))^{-1/p}$. The functions thus defined are connected by the following obvious relations:

$$f_m^n = -\sum_{j=1}^{n+1} g_{m,j}^n, \qquad f_j^{n+1} = -\sum_{m=1}^{n} g_{m,j}^n, \tag{3}$$

$$\sum_{m=1}^{n} f_m^n = -\sum_{m=1}^{n}\sum_{j=1}^{n+1} g_{m,j}^n = 1. \tag{4}$$

The sought-for series is obtained by writing after each element of the set F_n the corresponding *row* of the set G_n:

$$f_1^1 + g_{1,1}^1 + g_{1,2}^1 + f_1^2 + g_{1,1}^2 + g_{1,2}^2 + g_{1,3}^2 + f_2^2 + g_{2,1}^2 + g_{2,2}^2 + \cdots.$$

In what follows we shall write this series as $\sum_{n=1}^{\infty} h_n$.

THEOREM 3.2.1. *The series $\sum_{n=1}^{\infty} h_n$ converges to 0 in the metric of any of the spaces $L_p(Q)$, $1 \leq p < \infty$. There exists a permutation π such that the rearranged series $\sum_{n=1}^{\infty} h_{\pi(n)}$ converges to 1. The sum of any convergent rearrangement of the original series, $h_0 = \sum_{n=1}^{\infty} h_{\sigma(n)}$, is equal to either 0 or 1.*

PROOF. The convergence of our series to 0 is a consequence of relations (1)–(4). The rearranged series $\sum_{n=1}^{\infty} h_{\pi(n)}$ is constructed as follows: first write the element f_1^1, then after each element from F_n ($n > 1$) write the corresponding *column* of the set G_{n-1}:

$$f_1^1 + f_1^2 + g_{1,1}^1 + f_2^2 + g_{1,2}^1 + f_1^3 + g_{1,1}^2 + g_{2,1}^2 + f_2^3 + g_{1,2}^2 + g_{2,2}^2 + f_3^3 + \cdots.$$

This series converges to 1, thanks to the same relations (1)–(4).

Now let us turn to the last, main assertion of the theorem. It clearly suffices to consider the case of the weakest L_p-topology, i.e., the case $p = 1$. Notice that if $h_0 = \sum_{n=1}^{\infty} h_{\sigma(n)}$ is the sum of a convergent rearrangement, then, as seen from (3), for any $k \in \mathbf{N}$ all partial sums, starting with one of them, will not depend on the coordinate t_k. Since all terms take only integer values, h_0 is an integer constant. In view of this observations it remains to show that

$$\left\| h_0 - \frac{1}{2} \right\| \leq 1. \tag{$*$}$$

If $h_0 = 1$, then $(*)$ holds. Eliminating this case, we have $\|h_0 - 1\| \geq 1$. Fix an arbitrary $\delta > 0$. Take $K \in \mathbf{N}$ such that, for any $N \geq K$,

$$\left\| h_0 - \sum_{n=1}^{N} h_{\sigma(n)} \right\| \leq \delta, \tag{5}$$

and for any $m > l > K$,

$$\left\| \sum_{n=l}^{m} h_{\sigma(n)} \right\| \leq \delta. \tag{6}$$

Denote the partial sum $\sum_{n=1}^{K} h_{\sigma(n)}$ by h. In addition to the sets F_n and G_n introduced above, consider also the sets $V_n = \bigcup_{k=1}^{n}(F_k \cup G_k)$. Choose an index $M \in \mathbf{N}$ such that

$$h_{\sigma(n)} \in V_M \cup F_{M+1} \quad \text{for all } n \leq K.$$

Define the functions h_n^*, $\overline{h}_n$, h^* as follows:

$$h_n^* = \begin{cases} h_{\sigma(n)} & \text{if } h_{\sigma(n)} \in V_M \cup F_{M+1},\ n > K, \\ 0 & \text{otherwise,} \end{cases}$$

$$\overline{h}_n = \begin{cases} h_{\sigma(n)} & \text{if } h_{\sigma(n)} \in G_{M+1} \\ 0 & \text{otherwise,} \end{cases}$$

and $h^* = \sum_{N=K+1}^{\infty} h_n^*$. From (4) it follows that $h + h^* = 1$. Hence,

$$\|h^*\| = \|h - 1\| \geq \|h_0 - 1\| - \|h_0 - h\| \geq 1 - \delta. \tag{7}$$

Denote $n_0 = K$ and for $j = 0, 1, 2, 3$ set

$$n_{j+1} = \min\left\{ n \colon \frac{1}{4} - \frac{5\delta}{4} \leq \left\| \sum_{i=n_j+1}^{n} h_i^* \right\| \leq \frac{1}{4} - \frac{\delta}{4} \right\}. \tag{8}$$

The fact that the indices n_{j+1} are well defined follows from (6) and (7). For $j = 0, 1, 2, 3$ define the following functions:

$$h_{j+1}^{**} = \sum_{n=n_j+1}^{n_{j+1}} h_n^*, \quad \overline{\overline{h}}_{j+1} = \sum_{n=n_j+1}^{n_{j+1}} \overline{h}_n, \quad \hat{h}_{j+1} = \sum_{n=n_j+1}^{n_{j+1}} h_{\sigma(n)},$$

and for $j = 1, 2, 3, 4$ set $r_j = \hat{h}_j - h_j^{**} - \overline{\overline{h}}_j$. Let $h_5^{**} = \sum_{n=n_4+1}^{\infty} h_n^*$. Notice that r_j $(1 \leq j \leq 4)$ is the sum of all functions $h_{\sigma(n)}$ such that (a) $n_{j-1} < n \leq n_j$, and (b) $h_{\sigma(n)} \notin V_M \cup F_{M+1} \cup G_{M+1}$. The remaning part of the proof of inequality $(*)$ goes as follows: take the partial sum $H = \sum_{n=1}^{n_4} h_{\sigma(n)}$ of the series under consideration, write it in the form

$$H = h + \sum_{j=1}^{4} h_j^{**} + \sum_{j=1}^{4} \overline{\overline{h}}_j + \sum_{j=1}^{4} r_j,$$

and estimate $\|H - \frac{1}{2}\|$, and then $\|h_0 - \frac{1}{2}\|$. We shall need the following auxiliary assertion, the proof of which we omit.

LEMMA 3.2.1. *Let $(X, \mathcal{X}, \mu)$ and $(Y, \mathcal{Y}, \nu)$ be measure spaces with probability measures. Let $f(x,y)$ and $g(x,y)$ be functions in $L_1(X \times Y)$, each of which depends on only one variable: $f(x,y) = \tilde{f}(x)$, $g(x,y) = \tilde{g}(y)$. Then*

$$\|f + g\| \geq \|f\| + \|g\|[1 - 2(\mu \times \nu)(\operatorname{supp} f)],$$

where $\mu \times \nu$ is the product measure on $X \times Y$ and $\operatorname{supp} f$ denotes the support of the function f, i.e., the set of the points $(x, y) \in (X, Y)$ such that $f(x,y) \neq 0$. □

Returning now to the proof of the theorem, let us estimate $\|\overline{\overline{h}}_j\|$ from below. Applying Lemma 3.2.1 to the functions h_j^{**} and r_j (they depend on different coordinates), we obtain

$$\|h_j^{**} + r_j\| \geq \|h_j^{**}\| + \frac{1}{2}\|r_j\|, \quad 1 \leq j \leq 4 \tag{9}$$

(indeed, since the function h_j^{**} is integer valued, the condition $\|h_j^{**}\| \leq \frac{1}{4}$ implies $\operatorname{meas}(\operatorname{supp} h_j^{**}) \leq \frac{1}{4}$). On the other hand, from (6) is follows that $\|\hat{h}_j\| \leq \delta$. By (8) and (9),

$$\|\overline{\overline{h}}_j\| = \|h_j^{**} - r_j - \hat{h}_j\| \geq \|h_j^{**} + r_j\| - \|\hat{h}_j\| \geq \|h_j^{**}\| - \|\hat{h}_j\| \geq \frac{1}{4} - \frac{9\delta}{4}. \tag{10}$$

Next, let us estimate $\|h_5^{**}\|$ from above. Suppose $\|h_5^{**}\| > 11\delta$. Starting with this assumption, choose an index n_5 such that $\hat{h}_5^{**} = \sum_{n=n_4+1}^{n_5} h_n^*$ obeys the bounds $10\delta < \|\hat{h}_5^{**}\| \leq 11\delta$. Take $\overline{\overline{h}}_5 = \sum_{n=n_4+1}^{n_5} \overline{h}_n$. Similarly to estimate (10), we obtain $\|\overline{\overline{h}}_5\| \geq \|\hat{h}_5^{**}\| - \delta$. But $1 \geq \sum_{n=1}^{\infty} \|\overline{h}_n\| \geq \sum_{j=1}^{5} \|\overline{\overline{h}}_j\| > 1 - 9\delta + 9\delta = 1$. The contradiction we arrived at establishes the estimate $\|h_5^{**}\| \leq 11\delta$.

Consider now $\sum_{j=1}^{4} \|r_j\|$. We have

$$1 \geq \sum_{j=1}^{4} \|\overline{\overline{h}}_j\| = \sum_{j=1}^{4} \|\hat{h}_j - h_j^{**} - r_j\| \geq \sum_{j=1}^{4} \|h_j^{**} + r_j\| - \sum_{j=1}^{4} \|\hat{h}_j\|.$$

Combining (9), (10), and (8), we obtain

$$1 \geq 1 - 5\delta + \frac{1}{2}\sum_{j=1}^{4} \|r_j\| - 4\delta,$$

whence $\sum_{j=1}^{4} \|r_j\| \leq 18\delta$. Now let us estimate $\|H - \frac{1}{2}\|$. We have

$$\left\|H - \frac{1}{2}\right\| = \left\|h + \sum_{j=1}^{4} h_j^{**} + h_5^{**} + \sum_{j=1}^{4} \overline{\overline{h}}_j + \sum_{j=1}^{4} r_j - h_5^{**} - \frac{1}{2}\right\|$$

$$= \left\| h + h^* - \frac{1}{2} + \sum_{j=1}^{4} \overline{\overline{h}}_j + \sum_{j=1}^{4} r_j - h_5^{**} \right\|$$

$$\leq \left\| \frac{1}{2} + \sum_{j=1}^{4} \overline{\overline{h}}_j \right\| + \left\| \sum_{j=1}^{4} r_j \right\| + \|h_5^{**}\| \leq \frac{1}{2} + 18\delta + 11\delta = \frac{1}{2} + 29\delta.$$

Finally,

$$\left\| h_0 - \frac{1}{2} \right\| \leq \left\| H - \frac{1}{2} \right\| + \|h_0 - H\| \leq \frac{1}{2} + 29\delta + \delta = \frac{1}{2} + 30\delta,$$

which in view of the arbitrariness of δ completes the proof of the theorem. □

The proof given above belongs to K. Wozniakowski. P. A. Kornilov's proof uses a somewhat modified construction: essentially, a transition from Q to the segment $[0, 1]$ is effected by means of a measure-preserving mapping.

EXERCISE 3.2.1. Show that in a Hilbert space for any $n \in \mathbf{N}$ there exists a series whose sum range consists of exactly n points.

It is not known whether any set in a Hilbert space can serve as the sum range of a series. This problem is still open even for the case of finite sets.

DEFINITION 3.2.1. A point is said to be a *limit point* for the series $\sum_{k=1}^{\infty} x_k$ if it is the limit of some subsequence of the sequence of partial sums of some rearrangement of the series. The set of all such points, called the *limit-point range* of the series, will be denoted by $\mathrm{LPR}(\sum_{k=1}^{\infty} x_k)$.

Clearly, $\mathrm{LPR}(\sum_{k=1}^{\infty} x_k)$ is a closed set, and $\mathrm{SR}(\sum_{k=1}^{\infty} x_k) \subset \mathrm{LPR}(\sum_{k=1}^{\infty} x_k)$.

EXERCISE 3.2.2. Show that not every closed subset of a Hilbert space is the limit-point range of a series.

§3. Chobanyan's Theorem

In the two preceding sections we have demonstrated in a convincing manner that Steinitz's theorem does not carry over to the infinite-dimensional case. The relevant examples were constructed in L_p function spaces, but, as will be shown in Chapter 7, examples with similar properties can be constructed in any infinite-dimensional Banach space. We shall now adrress the question of what conditions imposed on the terms of a series in an infinite-dimensional space are sufficient to guarantee that the sum range is linear and has the properties stated in Steinitz's theorem.

We have already encountered an assertion of this kind, namely, Pecherskii's theorem (Theorem 2.3.1) which, as the reader undboubtedly recalls, holds true in the finite-dimensional as well as the infinite-dimensional case. The condition discussed in the present section represents the most elegant connection between the problems we are interested in and the theory of random series in Banach spaces. The proof will rely on Theorem 2.3.1, although Chobanyan's result [10] was obtained before that theorem.

From this moment on, $r_1, r_2, \dots$ will denote a sequence of independent random variables that take the values ± 1 with equal probabilities: $\mathbf{P}[r_i = 1] = \mathbf{P}[r_i = -1] = 1/2$. Let $x_1, x_2, \dots$ be elements of a Banach space X. Consider the random variables $S_k = \| \sum_{i=1}^k r_i x_i \|$.

LEMMA 3.3.1. *For any* $t > 0$,

$$\mathbf{P}\left[\sup_{k \leq n} S_k > t\right] \leq 2\mathbf{P}[S_n > t].$$

PROOF. Let Ω be the probability space on which the random variables r_i are defined. Define sets $A_i \subset \Omega$ as follows: $A_1 = [S_1 > t]$, $A_2 = [S_1 \leq t] \cap [S_2 > t], \dots, A_n = \bigcap_{i=1}^{n-1} [S_i \leq t] \cap [S_n > t]$. It is readily seen that

$$\mathbf{P}\left[\sup_{k \leq n} S_k > t\right] = \sum_{k=1}^{n} \mathbf{P}(A_k). \tag{2}$$

Denote $S_{n,k} = \| \sum_{i=1}^k r_i x_i - \sum_{i=k+1}^n r_i x_i \|$. By the triangle inequality, we have $S_k \leq \frac{1}{2}(S_n + S_{n,k})$. Consequently,

$$\mathbf{P}(A_k) \leq \mathbf{P}(A_k \cap [S_n > t]) + \mathbf{P}(A_k \cap [S_{n,k} > t]).$$

Since the probabilities in the right-hand side of this inequality coincide, $\mathbf{P}(A_k) \leq 2\mathbf{P}(A_k \cap [S_n > t])$. Using inequality (2) and the fact that the sets A_k are disjoint, we conclude that

$$\mathbf{P}\left[\sup_{k \leq n} S_k > t\right] \leq 2 \sum_{k=1}^{n} \mathbf{P}(A_k \cap [S_n > t]) \leq 2\mathbf{P}[S_n > t]. \qquad \square$$

Inequality (1) established above is a particular case of Lévy's inequality [92, Chapter 5, § 2]. An obvious consequence is the following inequality for mean values:

LEMMA 3.3.2. $\mathbf{E}\left(\sup_{k<n} S_k\right) \leq 2\mathbf{E}(S_n)$.

PROOF. It suffices to use the formula $\mathbf{E}(\nu) = \int_0^\infty \mathbf{P}[\nu > t]\, dt$ for the mean of a positive random variable and inequality (1). $\square$

LEMMA 3.3.3 (CHOBANYAN'S INEQUALITY). *Let $\{x_i\}_{i=1}^n$ be elements of a normed space, with $\sum_{i=1}^n x_i = 0$. Then there exists a permutation σ for which*

$$\sup_{k\le n}\left\|\sum_{i=1}^k x_{\sigma(i)}\right\| \le 2\,\mathbf{E}\left(\left\|\sum_{i=1}^n r_i x_i\right\|\right).$$

PROOF. Take for σ the permutation given by Lemma 2.3.4. Then

$$\sup_{k\le n}\left\|\sum_{i=1}^k x_{\sigma(i)}\right\| \le \mathbf{E}\left(\sup_{k\le n}\left\|\sum_{i=1}^k r_i x_{\sigma(i)}\right\|\right)$$

$$\le 2\,\mathbf{E}\left(\left\|\sum_{i=1}^n r_i x_{\sigma(i)}\right\|\right) \le 2\,\mathbf{E}\left(\left\|\sum_{i=1}^n r_i x_i\right\|\right). \qquad \square$$

THEOREM 3.3.1. *Let $\{x_i\}_{i=1}^\infty$ be elements of a Banach space with the property that $\lim_{m>n\to\infty}\mathbf{E}\left(\|\sum_{i=n}^m r_i x_i\|\right) = 0$. Then the assertion of Steinitz's theorem holds for the series $\sum_{i=1}^\infty x_i$.*

PROOF. Let $G = \{n, n+1, \dots, m\}$ and $A \subset G$. By the triangle inequality,

$$\mathbf{E}\left(\left\|\sum_{i=1}^m r_i x_i\right\|\right) \le \frac{1}{2}\left(\mathbf{E}\left(\left\|\sum_{i\in A} r_i x_i + \sum_{i\in G\setminus A} r_i x_i\right\|\right)\right.$$

$$\left.+\,\mathbf{E}\left(\left\|\sum_{i\in A} r_i x_i - \sum_{i\in G\setminus A} r_i x_i\right\|\right)\right) = \mathbf{E}\left(\left\|\sum_{i=1}^n r_i x_i\right\|\right).$$

Then, by Lemma 3.3.3 [resp. Lemma 2.3.3] condition (A) [resp. (B)] of Lemma 2.3.1 is satisfied. That is, analogues of the Permutation Lemma and the Rounding-off-Coefficients Lemma hold true. By Lemma 2.3.1, this implies the assertion of Steinitz's theorem. $\square$

COROLLARY 3.3.1 (CHOBANYAN'S THEOREM). *Let X be a Banach space. Then for the assertion of Steinitz's theorem to hold for a series $\sum_{n=1}^\infty x_n$ in X it is sufficient that the series $\sum_{n=1}^\infty r_n x_n$ converge almost everywhere, or, in other words, that the series $\sum_{n=1}^\infty \pm x_n$ converge for almost all choices of signs.*

PROOF. It is well known (see [92], Chapter 5, § 5) that the condition

$$\lim_{m>n\to\infty}\mathbf{E}\left\|\sum_{i=n}^m r_i x_i\right\| = 0$$

is equivalent to the almost everywhere convergence of the series $\sum_{n=1}^\infty r_n x_n$.

EXERCISE 3.3.1. Prove directly that the condition of almost everywhere convergence of the series $\sum_{n=1}^{\infty} r_n x_n$ is preserved for any rearrangement of its terms. Show that Corollary 3.3.1 follows from this conclusion and Pecherskii's theorem.

REMARK 3.3.1. S. A. Chobanyan also proved (see [95]) that under the conditions of Lemma 3.3.3 the two mean values

$$E_1 = \frac{1}{n!} \sum_{\pi} \sup_{k \leq n} \left\| \sum_{i=1}^{k} x_{\pi(i)} \right\|$$

and

$$E_2 = \mathbf{E} \left(\left\| \sum_{i=1}^{n} r_i x_i \right\| \right)$$

are equivalent, i.e., $C_1 E_1 \leq E_2 \leq C_2 E_1$, where C_1 and C_2 are constants independent of $\{x_i\}$ and n.

§4. The Khinchin Inequalities and the Theorem of M. I. Kadets on Conditionally Convergent Series in L_p

Conditions for the linearity of the sum range of a conditionally convergent series in an infinite-dimensional space were obtained for the first time in 1954 by M. I. Kadets [31] for the case of the L_p spaces. Various generalizations and sharpened forms of these results were proved by E. M. Nikishin [62], S. L. Troyanskii [90], V. P. Fonf [23], D. V. Pecherskii [67], V. M. Kadets [35], M. I. Ostrovskii [65], and others. The culmination of these generalizations are the theorems of Pecherskii and of Chobanyan, already familiar to the reader. A large part of the preceding results can be be deduced from these, strongest theorems. Here we shall use Chobanyan's theorem to establish a sufficient condition, due to M. I. Kadets, for the linearity of the sum range of a series in L_p. This result, which historically was the first in this direction, asserts that, in a space L_p, if $1 < p < 2$, then for the set $\mathrm{SR}(\sum_{k=1}^{\infty} x_k)$ to be linear it suffices that $\sum_{k=1}^{\infty} \|x_k\|^p < \infty$ (condition of p-absolute convergence of the series), while if $p \geq 2$, $p \neq \infty$ it suffices that $\sum_{k=1}^{\infty} \|x_k\|^2 < \infty$. For the proof we need some special inequalities, known as the Khinchin inequalities, the exposition of which is our next objective.

Let $x_1, \ldots, x_n$ be arbitrary fixed real numbers. Consider the quantity $M_p = \left(\mathbf{E}|\sum_{i=1}^{n} r_i x_i|^p\right)^{1/p}$, where r_i are independent random variables that take the values ± 1 with equal probabilities; M_p is the p-mean value of the 2^n combinations of the form $|\alpha_1 x_1 + \cdots + \alpha_n x_n|$, where the coefficients α_i take the values ± 1. If we put $p = 2$ and then open the parentheses and reduce the like terms in the expression

$$M_2 = \left[\frac{1}{2^n} \sum_{\alpha_i = \pm 1} \left(\sum_{i=1}^{n} \alpha_i x_i \right)^2 \right]^{1/2},$$

we see that $M_2 = \left(\sum_{i=1}^n |x_i|^2\right)^{1/2}$. It directly follows from the properties of p-means that M_p does not decrease with the growth of p: $M_1 \le M_p \le M_q$ for $1 \le p \le q < \infty$. By the *Khinchin inequalities* one usually means the inequalities

$$a_p \left(\sum_{i=1}^n |x_i|^2\right)^{1/2} \le M_p \le \left(\sum_{i=1}^n |x_i|^2\right)^{1/2}, \quad 1 \le p \le 2, \tag{1}$$

$$\left(\sum_{i=1}^n |x_i|^2\right)^{1/2} \le M_p \le A_p \left(\sum_{i=1}^n |x_i|^2\right)^{1/2}, \quad 2 \le p < \infty, \tag{2}$$

where the coefficients $0 < a_p < 1 \le A_p < \infty$ depend on p, but not on n or $x_1, \dots, x_n$.

PROOF OF THE KHINCHIN INEQUALITIES. THE CASE $1 \le p \le 2$. The right inequality is a direct consequence of the relation $M_p \le M_2$ and the expression of M_2. Since $M_1 \le M_p$, it suffices to establish the left inequality for M_1. Thus, we have to show that

$$a \left(\sum_{i=1}^n |x_i|^2\right)^{1/2} \le \mathbf{E}\left(\left|\sum_{i=1}^n r_i x_i\right|\right) \tag{3}$$

for some $a > 0$ that does not depend on x_i or on n. Let us divide both sides of inequality (3) by $\left(\sum_{i=1}^n |x_i|^1\right)^{1/2}$ and denote $t_i = x_i / \left(\sum_{i=1}^n |x_i|^1\right)^{1/2}$. Then (3) can be recast as

$$\mathbf{E}\left(\left|\sum_{i=1}^n r_i t_i\right|\right) \ge a > 0, \quad \sum_{i=1}^n t_i^2 = 1. \tag{4}$$

Using the inequality $|t| \ge 1 - \cos t,\ t \in \mathbf{R}$, and the identity

$$\sum_{\alpha_i = \pm 1} \cos\left(\sum_{i=1}^n \alpha_i t_i\right) = 2^n \prod_{i=1}^n \cos t_i,$$

we conclude that

$$\mathbf{E}\left(\left|\sum_{i=1}^n r_i t_i\right|\right) = 2^{-n} \sum_{\alpha_i = \pm 1} \left|\sum_{i=1}^n \alpha_i t_i\right|$$

$$\ge 2^{-n} \sum_{\alpha_i = \pm 1} \left(1 - \cos\left(\sum_{i=1}^n \alpha_i t_i\right)\right) = 1 - \prod_{i=1}^n \cos t_i.$$

To prove inequality (4) it remains to show that

$$\min\left\{1 - \prod_{i=1}^n \cos t_i : \sum_{i=1}^n t_i^2 = 1\right\}$$

is bounded from below by a positive number that does not depend on n. Omitting the details, let us observe that the minimum in question is attained for $t_1 = \cdots = t_n = n^{-1/2}$ and is equal to $1 - \left(\cos n^{-1/2}\right)^n$, which in turn is estimated from below by $1 - e^{-1/2}$. Thus, Khinchin's inequality is established for $1 \leq p \leq 2$, with a constant $a_p \geq 1 - e^{-1/2}$.

THE CASE $2 < p < \infty$. The left inequality follows from the fact that M_p does not decrease when p is increased. To prove the right inequality let us again divide both sides by $\left(\sum_{i=1}^n |x_i|^2\right)^{1/2}$ and denote $t_i = x_i / \left(\sum_{i=1}^n |x_i|^2\right)^{1/2}$. Then the inequality becomes

$$\left(\mathbf{E}\left(\left|\sum_{i=1}^n r_i t_i\right|^p\right)\right)^{1/p} \leq A_p, \quad \sum_{i=1}^n t_i^2 = 1. \tag{5}$$

Here we use the inequality

$$|t|^p \leq C_p(\cosh t - 1), \quad t \in \mathbf{R},$$

which holds for any $p \geq 2$, with the constant C_p depending on p. Performing the corresponding substitution in the left-hand side of inequality (5) we obtain, after simplifications,

$$\mathbf{E}\left(\left|\sum_{i=1}^n r_i t_i\right|^p\right) \leq C_p\left(\prod_{i=1}^n \cosh t_i - 1\right).$$

As in the preceding case,

$$\max\left\{\prod_{i=1}^n \cosh t_i - 1 \colon \sum_{i=1}^n t_i^2 = 1\right\}$$

is attained for $t_1 = \cdots = t_n = n^{-1/2}$ and is equal to $\left(\cosh n^{-1/2}\right)^n - 1$, which in turn is bounded above by $e^{1/2} - 1$. This completes the proof of the Khinchin inequality for the case $2 \leq p < \infty$. □

The sharp values of the constants a_p and A_p are presently known (see [70], [88], [94]). In particular, $a_1 = 1/\sqrt{2}$. It is possible to introduce the quantities $M_p = \left(\mathbf{E}\left\|\sum_{i=1}^n r_i x_i\right\|^p\right)^{1/p}$ for x_i belonging to an arbitrary normed space (an not only for $x_i \in \mathbf{R}$). The surprizing inequality of Kahane (see [99]) shows that in this case the p-mean values M_p with different p's are also equivalent.

DEFINITION 3.4.1. A normed space X is said to have *type p with constant C* if, for any finite set $\{x_i\}_{i=1}^n$ of elements of X the following inequality holds:

$$\mathbf{E}\left(\left\|\sum_{i=1}^n r_i x_i\right\|\right) \leq C\left(\sum_{i=1}^n \|x_i\|^p\right)^{1/p}.$$

By the triangle inequality, any space has type $p = 1$. Hence, of interest are only the spaces of type $p > 1$. We shall discuss such spaces in more detail in Chapter 5; here we need only their definition.

THEOREM 3.4.1 (S. A. CHOBANYAN). *Let the space X have type p and let the series $\sum_{k=1}^{\infty} x_k$ be such that $\sum_{k=1}^{\infty} \|x_k\|^p < \infty$. Then the assertion of Steinitz's theorem holds for* $\mathrm{SR}(\sum_{k=1}^{\infty} x_k)$.

PROOF. This is an immediate consequence of Theorem 3.3.1. □

To obtain the theorem of M. I. Kadets it remains to deduce from the Khinchin inequalities that L_p spaces have a type.

THEOREM 3.4.2. *The spaces $L_p(\Omega, \mu)$ with $2 < p < \infty$ have type 2.*

PROOF. Let $f_1, \ldots, f_n \in L_p(\Omega, \mu)$. Then

$$\mathbf{E}\left(\left\|\sum_{i=1}^{n} r_i f_i\right\|\right) \le \left(\mathbf{E}\left(\left\|\sum_{i=1}^{n} r_i f_i\right\|^p\right)\right)^{1/p} = \left(\mathbf{E}\int_\Omega \left|\sum_{i=1}^{n} r_i f_i(t)\right|^p d\mu\right)^{1/p} =$$

$$= \left(\int_\Omega \mathbf{E}\left|\sum_{i=1}^{n} r_i f_i(t)\right|^p d\mu\right)^{1/p} \le A_p \left(\int_\Omega \left(\sum_{i=1}^{n} |f_i(t)|^2\right)^{p/2} d\mu\right)^{1/p}$$

(in the last step we used the Khinchin inequality). If one regards the functions $|f_i(t)|^2$ as elements of the space $L_{p/2}(\Omega, \mu)$, then the triangle inequality gives

$$\left(\int_\Omega \left(\sum_{i=1}^{n} |f_i(t)|^2\right)^{p/2} d\mu\right)^{1/p} \le \sum_{i=1}^{n} \left(\int_\Omega (|f_i(t)|^2)^{p/2} d\mu\right)^{1/p} = \sum_{i=1}^{n} \|f_i\|_{L_p}^2.$$

Consequently,

$$\mathbf{E}\left(\left\|\sum_{i=1}^{n} r_i f_i\right\|\right) \le A_p \left(\sum_{i=1}^{n} \|f_i\|^2\right)^{1/2}.$$ □

THEOREM 3.4.3. *The spaces $L_p(\Omega, \mu)$ with $1 \le p \le 2$ have type p.*

PROOF. As in the preceding assertion, the Khinchin inequality yields

$$\mathbf{E}\left(\left\|\sum_{i=1}^{n} r_i f_i\right\|\right) \le \left(\int_\Omega \left(\sum_{i=1}^{n} |f_i|^2\right)^{p/2} d\mu\right)^{1/p}.$$

Using the inequality

$$\left(\sum |c_i|^p\right)^{1/p} \ge \left(\sum |c_i|^2\right)^{1/2},$$

which expresses the monotone dependence of the l_p-norm on p, we obtain

$$\mathbf{E}\left(\left\|\sum_{i=1}^{n} r_i f_i\right\|\right) \le \left(\int_\Omega \sum_{i=1}^{n} |f_i|^p d\mu\right)^{1/p} = \left(\sum_{i=1}^{n} \|f_i\|^p\right)^{1/p}.$$ □

COROLLARY 3.4.1 (M. I. KADETS' THEOREM). *Let $1 \leq p < \infty$ and $r = \min\{2, p\}$. Then in order for the sum range of a series $\sum_{k=1}^{\infty} x_k$ in L_p to be linear it suffices that $\sum_{k=1}^{\infty} \|x_k\|^r < \infty$.* □

REMARK 3.4.1. In the preceding assertion $\mathrm{SR}(\sum_{k=1}^{\infty} x_k)$ has all the properties stated in Steinitz's theorem, i.e., not only the set $\mathrm{SR}(\sum_{k=1}^{\infty} x_k)$ is linear, but it also has the form $x_0 + \Gamma^{\perp}$; this fact was first observed by V. P. Fonf in 1972 [23]. More precisely, Fonf's theorem gives a sufficient condition for the equality $\mathrm{SR}(\sum_{k=1}^{\infty} x_k) = x_0 + \Gamma^{\perp}$ to hold for series in uniformly smooth spaces: the condition is that $\sum_{k=1}^{\infty} \rho(\|x_k\|) < \infty$, where $\rho(t)$ is the modulus of smoothness of the space (for the definition of this notion see [13]).

EXERCISE 3.4.1. Carry out the calculations needed to solve the extremum problems involved in the proof of the Khinchin inequalities.

EXERCISE 3.4.2. Show that theorems 3.4.1, 3.4.2, and 3.4.3 are sharp, i.e., the exponents appearing in these theorems cannot be improved.

REMARK 3.4.2. An analysis of the foregoing arguments yields the following, weaker sufficient condition: for the equality $\mathrm{SR}(\sum_{k=1}^{\infty} x_k) = x_0 + \Gamma^{\perp}$ to hold in an L_p space it is sufficient that $g(t) = \left(\sum_{k=1}^{\infty} |x_k(t)|^2\right)^{1/2} \in L_p$. This result, which is stronger than M. I. Kadets' theorem, belongs to Nikishin [62], and has a lattice-theoretic, rather than a Banach space-theoretic character.

EXERCISE 3.4.3. Verify that none of the sufficient conditions for the linearity of the sum range given above is necessary.

CHAPTER 4
UNCONDITIONALLY CONVERGENT SERIES

Starting with this chapter we move deeper and deeper into the realm of the concepts, repesentations, and methods of the contemporary theory of Banach spaces.

§1. The Dvoretzky-Rogers Theorem

In this section we shall establish the following general fact: in any infinite-dimensional Banach space there are series that converge unconditionally, but not absolutely. Thus, the equivalence between unconditional and absolute convergence characterizes the finite-dimensional spaces within the class of all Banach spaces.

LEMMA 4.1.1 (THE DVORETZKY-ROGERS LEMMA). *Let X be an n-dimensional normed space. Then there exist elements $\{x_i\}_{i=1}^n \subset S(X)$ such that*

$$\left\|\sum_{i=1}^m t_i x_i\right\| \leq \left(1+\sqrt{\frac{m(m-1)}{n}}\right)\cdot\sqrt{\sum_{i=1}^m t_i^2} \tag{1}$$

for any scalars $\{t_i\}_{i=1}^m$ and any m, $1 \leq m \leq n$.

PROOF. In the unit ball $U(X)$ of the space X pick an ellipsoid E of maximal volume. (Recall that in an n-dimensional linear space an ellipsoid is defined as the set of all points for which the value of some positive-definite quadratic form, which defines the ellipsoid, is smaller than or equal to 1.) Equip X with the inner product generated by the ellipsoid E (or, equivalently, by the quadratic form that defines E). Denote by $\|\cdot\|_E$ the corresponding norm, with respect to which E is the unit ball of X. Obviously, $\|x\| \leq \|x\|_E$ for all $x \in X$. Since the notion of volume in a finite-dimensional space is defined up to a constant factor, we may assume that the volume of E is equal to one. Then the volume of any ellipsoid will be the product of the lengths of its principal semi-axes, calculated in the metric $\|\cdot\|_E$.

Let us construct, by induction on m, an orthonormal (with respect to the inner product introduced above) basis $\{u_i\}_{i=1}^n$ and a set of vectors $\{x_i\}_{i=1}^n$, with the following properties:

(A) $\|x_m\| = \|x_m\|_E = 1,\ 1 \le m \le n$;
(B) $x_m = \sum_{j=1}^{m} a_{m,j} u_j,\ a_{m,m} > 0$;
(C) $a_{m,1}^2 + a_{m,2}^2 + \cdots + a_{m,m-1}^2 = 1 - a_{m,m}^2 \le \frac{m-1}{n}$.

Take for $u_1 = x_1$ an arbitrary point at which the ellipsoid E touches the unit ball $U(X)$, i.e., a point a such that $\|a\| = \|a\|_E = 1$. The existence of such points on the unit sphere $S(X)$ is guaranteed by the finite-dimensionality of X and the maximality of E. Suppose we have already found vectors $\{u_i\}_{i=1}^n$ and $\{x_i\}_{i=1}^n$ satisfying the requirements

(A_m) $\|x_i\| = \|x_i\|_E = 1,\ 1 \le i \le m$;
(B_m) $x_i = \sum_{j=1}^{i} a_{i,j} u_j,\ a_{i,i} > 0$;
(C_m) $a_{i,1}^2 + a_{i,2}^2 + \cdots + a_{i,i-1}^2 = 1 - a_{i,i}^2 \le \frac{i-1}{n}$.

Complete the set $\{u_i\}_{i=1}^n$ in an arbitrary way to an orthonormal basis by adding vectors $v_{m+1}, \dots, v_n$. Now let us carry out the next inductive step. Given an arbitrary $\varepsilon > 0$, consider the perturbed ellipsoid E_ε formed by the points $x = (g_1, \dots, g_n)$ whose coordinates in the orthonormal basis constructed above satisfy the inequality

$$(1+\varepsilon)^{n-m}(g_1^2 + \cdots + g_m^2) + (1+\varepsilon+\varepsilon^2)^{-m}(g_{m+1}^2 + \cdots + g_n^2) \le 1. \tag{2}$$

The volume V_ε of E_ε is larger than the volume of the basic ellipsoid E:

$$V_\varepsilon = \frac{(1+\varepsilon+\varepsilon^2)^{m(n-m)}}{(1+\varepsilon)^{(n-m)\cdot m}} > 1.$$

Consequently, one can find a point $x(\varepsilon) \in E_\varepsilon$ such that $x(\varepsilon) \in S(X)$ (even more, one can find points in E_ε that lie outside $U(X)$). Since $\|x(\varepsilon)\|_E \ge \|x(\varepsilon)\| = 1$, for this point we have $g_1^2 + \cdots + g_n^2 \ge 1$. Substracting the last inequality from (2), we see that the coordinates of $x(\varepsilon)$ satisfy the inequality

$$\left[(1+\varepsilon)^{n-m} - 1\right](g_1^2 + \cdots + g_m^2) + \left[(1+\varepsilon+\varepsilon^2)^{-m} - 1\right](g_{m+1}^2 + \cdots + g_n^2) \le 0. \tag{3}$$

Let us divide both sides of (3) by ε and pass to the limit for a sequence of numbers $\varepsilon \searrow 0$ for which the corresponding sequence of vectors $x(\varepsilon)$ converges to some vector $y \in S(X)$. This yields the following inequality for the coordinates of y:

$$(n-m)(g_1^2 + \cdots + g_m^2) - m(g_{m+1}^2 + \cdots + g_n^2) \le 0. \tag{4}$$

Now set $x_{m+1} = y$. Since $x_{m+1} \in S(X) \cap E$, $\|x_{m+1}\| = \|x_{m+1}\|_E = 1$. Notice that from (4) it follows that for $m < n$ the vector x_{m+1} is linearly independent with the vectors $u_1, \dots, u_m$. Now choose a vector $u_{m+1} \in S(X)$ in the linear span of $u_1, \dots, u_m, x_{m+1}$ so that u_{m+1} is orthogonal to $u_1, \dots, u_m$ and the coefficient $a_{m+1,m+1}$ in the decomposition

$$x_{m+1} = \sum_{j=1}^{m+1} a_{m+1,j} u_j$$

is positive. Then the vectors x_{m+1} and u_{m+1} satisfy conditions (A_{m+1}) and (B_{m+1}). Let us verify that condition (C_{m+1}) is also satisfied.

By construction, the numbers g_i figuring in inequality (4) are connected with the numbers $a_{m+1,j}$ by the relations $a_{m+1,j} = g_j$ for $j \leq m$, and $a^2_{m+1,m+1} = \sum_{i=1}^{n} g_i^2$. Consequently, we can rewrite (4) in the form

$$(n-m)\big(a^2_{m+1,1} + \cdots + a^2_{m+1,m}\big) - m a^2_{m+1,m+1} \leq 0,$$

or, equivalently,

$$n\big(a^2_{m+1,1} + \cdots + a^2_{m+1,m}\big) - m\big(a^2_{m+1,1} + \cdots + a^2_{m+1,m+1}\big) \leq 0.$$

In conjunction with the relation $\|x_{m+1}\|_E^2 = \sum_{j=1}^{m+1} a^2_{m,j} = 1$, this yields the required condition

$$a^2_{m+1,1} + \cdots + a^2_{m+1,m} = 1 - a^2_{m+1,m+1} \leq \frac{m}{n}.$$

Thus, by completing the induction process we produce an orthonormal set $\{u_i\}_{i=1}^n$ and a set $\{x_i\}_{i=1}^n$ of norm-one elements, which satisfy the conditions (A), (B), and (C). We claim that the set $\{x_i\}_{i=1}^n$ satisfies inequality (1). Indeed,

$$\left\|\sum_{i=1}^m t_i x_i\right\| \leq \left\|\sum_{i=1}^m t_i x_i\right\|_E \leq \left\|\sum_{i=1}^m t_i u_i\right\|_E + \left\|\sum_{i=1}^m t_i(u_i - x_i)\right\|_E$$

$$\leq \sqrt{\sum_{i=1}^m t_i^2} + \sum_{i=1}^m |t_i| \cdot \|u_i - x_i\|_E \leq \sqrt{\sum_{i=1}^m t_i^2}\left(1 + \sqrt{\sum_{i=1}^m \|u_i - x_i\|_E^2}\right). \tag{5}$$

The norm $\|u_i - x_i\|_E$ is estimated by

$$\|u_i - x_i\|_E^2 = a^2_{i,1} + a^2_{i,2} + \cdots + a^2_{i,i-1} + (1 - a_{i,i})^2$$

$$= (1 - a^2_{i,i}) + (1 - a_{i,i})^2 \leq 2(1 - a^2_{i,i}) \leq \frac{2(i-1)}{n}. \tag{6}$$

Here we made use of all the conditions (A), (B), (C), including the requirement that $a_{i,i} > 0$. Substituting (6) in (5), we obtain the needed inequality (1). □

COROLLARY 4.1.1. *Let $n \geq m(m-1)$, and let X be an arbitrary n-dimensional normed space. Then on the unit sphere $S(X)$ one can find elements $\{x_i\}_{i=1}^m$ with the property that*

$$\left\|\sum_{i=1}^m t_i x_i\right\| \leq 2\sqrt{\sum_{i=1}^m t_i^2}$$

for any choice of coefficients $\{t_i\}_{i=1}^m$. □

TEOREM 4.1.1 (THE DVORETZKY-ROGERS THEOREM). *Let X be an infinite-dimensional Banach space. Then for any sequence $\{a_i\}_{i=1}^{\infty}$ of positive numbers satisfying the condition $\sum_{i=1}^{\infty} a_i^2 < \infty$ there exists an unconditionally convergent series $\sum_{i=1}^{\infty} x_i$ in X for which $\|x_i\| = a_i$, $i = 1, 2, \ldots$.*

PROOF. Divide the sequence $\{a_i\}_{i=1}^{\infty}$ into segments $\{a_i\}_{i=m_j+1}^{m_{j+1}}$, $0 = m_0 < m_1 < m_2 < \ldots$, such that

$$\sum_{i=m_j+1}^{m_{j+1}} a_i^2 \leq 2^{-2j}, \quad j = 1, 2, \ldots.$$

Choose the first m_1 vectors x_i in arbitrary way, subject only to the requirement that $\|x_i\| = a_i$, $1 \leq i \leq m_1$. Next, take a sequence of subspaces X_j of X of dimensions $\dim X_j = (m_{j+1} - m_j)(m_{j+1} - m_j - 1)$. By Corollary 4.1.1, in each X_j one can find a set of norm-one elements $\{\overline{x}_i\}_{i=m_j+1}^{m_{j+1}}$ for which

$$\left\| \sum_{i=m_j+1}^{m_{j+1}} t_i \overline{x}_i \right\| \leq 2 \left(\sum_{i=m_j+1}^{m_{j+1}} t_i^2 \right)^{1/2}$$

for any choice of coefficients t_i. Set $x_i = a_i \overline{x}_i$. For the choice $|t_i| = a_i$ the preceding inequality yields

$$\max_{\alpha_i = \pm 1} \left\| \sum_{i=m_j+1}^{m_{j+1}} \alpha_i x_i \right\| \leq 2 \left(\sum_{i=m_j+1}^{m_{j+1}} a_i^2 \right)^{1/2}. \tag{7}$$

The sequence $\{x_i\}_{i=1}^{\infty}$ constructed in the indicated manner satisfies the condition $\|x_i\| = a_i$, $i = 1, 2, \ldots$. It remains to show that the series $\sum_{i=1}^{\infty} x_i$ is unconditionally convergent. To this end let us take an arbitrary segment $\sum_{m=1}^{n} x_i$ of the series and, using inequality (7), let us estimate from above the norm $\|\sum_{i=m+1}^{n} \alpha_i x_i\|$. Let j be a number such that $m_j \leq m \leq m_{j+1}$. Then

$$\left\| \sum_{i=m+1}^{n} \alpha_i x_i \right\| \leq \sum_{k=j}^{\infty} \max_{\alpha_i = \pm 1} \left\| \sum_{i=m_j+1}^{m_{j+1}} \alpha_i x_i \right\| \leq \sum_{k=j}^{\infty} 2^{-k+1} = 2^{-j+2}$$

(here we used the fact that if $A \subset B \subset \mathbf{N}$, then

$$\max_{\alpha_i = \pm 1} \left\| \sum_{i \in A} \alpha_i x_i \right\| \leq \max_{\alpha_i = \pm 1} \left\| \sum_{i \in B} \alpha_i x_i \right\|,$$

which in turn is an immediate consequence of the triangle inequality). Since j grows when m does, according to Cauchy's criterion our series diverges for any arrangement of signs $\alpha_i = \pm 1$, i.e., unconditionally. □

Since in the formulation of the last theorem the sequence $\{a_i\}_{i=1}^\infty$ does not necessarily satisfy the condition $\sum_{i=1}^\infty a_i < \infty$, the Dvoretzky-Rogers theorem establishes, in particular, the existence, in any infinite-dimensional Banach space, of unconditionally convergent series that are not absolutely convergent. A proof of this fact based on the theory of absolutely summing operators (see Definition 4.3.1) can be found in [14].

EXERCISE 4.1.1. Investigate the possibility of choosing elements $\{x_i\}_{i=1}^m$ in any n-dimensional Banach space X, such that $\|x_i\| = 1$, $i = 1, \ldots, m$, and

$$\left\| \sum_{i=1}^m t_i x_i \right\| \geq \frac{1}{2} \sqrt{\sum_{i=1}^m t_i^2}$$

for all collections $\{t_i\}_{i=1}^m$. Find the relation between m and n.

§2. Orlicz's Theorem on Unconditionally Convergent Series in L_p Spaces

As we already know, in finite-dimensional spaces unconditional convergence implies absolute convergence (Theorem 1.3.3). In infinite-dimensional Banach spaces this is no longer the case: by the Dvoretzky-Rogers theorem, in each such space there is a series that converges unconditionally, but not absolutely. One can then ask whether unconditional convergence of a series of elements in a Banach space imposes any constraints whatsoever on the behavior of the numerical series formed by the norms of its terms. A trivial constraint, which holds not only for unconditionally convergent, but also for ordinarily convergent series, is that the general term must tend to zero. Modifying the construction given in Example 1.3.1 one can readily see that in the space c_0 the norms of the terms of an unconditionally convergent series are not requied to satisfy any nontrivial conditions. However, there are spaces whose nature imposes rather strong constraints on the behavior of unconditionally convergent series. The first result in this direction was obtained by W. Orlicz in 1930.

THEOREM 4.2.1. *Let the series $\sum_{i=1}^\infty x_i$ of elements of a space L_p converge unconditionally. Then for $1 \leq p \leq 2$ the series $\sum_{i=1}^\infty \|x_i\|^2$ converges, and for $2 \leq p < \infty$ the series $\sum_{i=1}^\infty \|x_i\|^p$ converges.*

We will reduce the proof of Orlicz's theorem to the proof of an assertion on the sum of a finite number of elements of a Banach space.

DEFINITION 4.2.1. The space X is said to have the *Orlicz property with exponent* r if for any unconditionally convergent series $\sum_{i=1}^\infty x_i$ of elements of X the series $\sum_{i=1}^\infty \|x_i\|^r$ converges.

DEFINITION 4.2.2. The space X is said to have *M-cotype r with constant* $\gamma > 0$ if the inequality

$$\max\left\{\left\|\sum \alpha_i x_i\right\| : \alpha_i = \pm 1\right\} \geq \gamma \left(\sum \|x_i\|^r\right)^{1/r} \tag{1}$$

holds for any finite set $\{x_i\}$ of elements of X.

THEOREM 4.2.2. *The following assertions are equivalent:*
(A) *the space X has the Orlicz property with exponent r;*
(B) *the space X has M-cotype r (i.e., there exists a $\gamma > 0$ such that X has M-cotype r with constant γ).*

PROOF. First let us prove the implication (A) $\Rightarrow$ (B). Suppose (B) does not hold. Then there exist a sequence $\{x_i\}_{i=n_k+1}^{n_{k+1}}$ of finite subsets of the space X, $1 = n_0 < n_1 < n_2 < \dots$, for which

$$\sum_{i=n_k+1}^{n_{k+1}} \|x_i\|^r \geq 1, \quad \max_{\alpha_i=\pm 1} \left\| \sum_{i=n_k+1}^{n_{k+1}} \alpha_i x_i \right\| \leq 2^{-k}.$$

The series $\sum_{i=1}^{\infty} x_i$ formed by these elements converges unconditionally, but $\sum_{i=1}^{\infty} \|x_i\|^r = \infty$. Thus, negation of (B) implies negation of (A), which is equivalent to (A) $\Rightarrow$ (B).

Now let us show that (B) $\Rightarrow$ (A). Suppose there exists an unconditionally convergent series $\sum_{i=1}^{\infty} x_i$ for which $\sum_{i=1}^{\infty} \|x_i\|^r = \infty$. From Gel'fand's theorem (Theorem 1.3.4) it follows that the sums $\sum_{i=m+1}^{n} \alpha_i x_i$ are uniformly bounded, i.e.,

$$\max \left\{ \sum_{i=m+1}^{n} \|\alpha_i x_i\| : \alpha_i = \pm 1, \ m, n \in \mathbf{N} \right\} \leq C < \infty.$$

On the other hand, from the fact that $\sum_{i=1}^{\infty} \|x_i\|^r = \infty$ it follows that there exists a sequence of segments of the series such that

$$\sum_{i=m_k+1}^{n_k} \|x_i\|^r \to \infty, \quad 1 \leq m_1 < n_1 < m_2 < n_2 < \dots.$$

This clearly contradicts inequality (1), i.e., (B) does not hold. This establishes the implication (B) $\Rightarrow$ (A). □

REMARK 4.2.1. The finite-dimensional spaces have M-cotype $r = 1$, whereas by the Dvoretzky-Rogers theorem, in the infinite-dimensional case one necessarily has $r \geq 2$.

To complete the proof of Theorem 4.2.1 it remains to show that the spaces L_p have the corresponding M-cotype. This fact will be derived from the Khinchin inequalities.

PROOF OF ORLICZ'S THEOREM. Consider first the case $1 \leq p \leq 2$. Let us pick a finite collection $\{x_i\}_{i=1}^{n}$ of elements in $L_p(\Omega, \mu)$ and, in accordance with Theorem 4.2.2, let us estimate from below the quantity

$$\max_{\alpha_i=\pm 1} \left\| \sum_{i=1}^{n} \alpha_i x_i \right\|^p = \max_{\alpha_i=\pm 1} \int_{\Omega} \left| \sum_{i=1}^{n} \alpha_i x_i(t) \right|^p d\mu(t).$$

To this end let us first replace the maximum by the mean value over the $\alpha_i = \pm 1$ and then apply the Khinchin inequality:

$$\max_{\alpha_i=\pm 1} \left\| \sum_{i=1}^{n} \alpha_i x_i \right\|^p \geq \mathbf{E} \int_\Omega \left| \sum_{i=1}^{n} r_i x_i(t) \right|^p d\mu(t)$$

$$\int_\Omega \mathbf{E} \left| \sum_{i=1}^{n} r_i x_i(t) \right|^p d\mu(t) \geq (a_p)^p \int_\Omega \left(\sum_{i=1}^{n} |x_i(t)|^2 \right)^{p/2} d\mu(t).$$

The integrand in the last integral can be regarded as the norm of the value at the point t of the vector-function

$$y(t) = \{|x_1(t)|^p, \ldots, |x_n(t)|^p\}$$

in the coordinate space $l_{2/p}^{(n)}$:

$$\left(\sum_{i=1}^{n} |x_i(t)|^2 \right)^{p/2} = \left(\sum_{i=1}^{n} (|x_i(t)|^p)^{2/p} \right)^{p/2} = \|y(t)\|_{2/p}.$$

By the triangle inequality, the integral of the norm is not smaller than the norm of the integral:

$$\int_\Omega \left(\sum_{i=1}^{n} |x_i(t)|^2 \right)^{p/2} d\mu(t) = \int_\Omega \|y(t)\|_{2/p} d\mu(t) \geq \left\| \int_\Omega y(t) d\mu(t) \right\|_{2/p}$$

$$= \left(\sum_{i=1}^{n} \left(\int_\Omega |x_i(t)|^p d\mu(t) \right)^{2/p} \right)^{p/2} = \left(\sum_{i=1}^{n} \|x_i\|_{L_p}^2 \right)^{p/2}.$$

This finally yields the inequality

$$\max_{\alpha_i=\pm 1} \left\| \sum_{i=1}^{n} \alpha_i x_i \right\|^p \geq a_p \left(\sum_{i=1}^{n} \|x_i\|^2 \right)^{1/2},$$

which, by Theorem 4.2.2, proves Orlicz's theorem for the case $1 \leq p \leq 2$.

Now let us consider the case $2 \leq p < \infty$. As in the first part of the proof, replacing the maximum by the mean value and applying the Khinchin inequality we obtain

$$\max_{\alpha_i=\pm 1} \left\| \sum_{i=1}^{n} \alpha_i x_i \right\|^p \geq \int_\Omega \left(\sum_{i=1}^{n} |x_i(t)|^2 \right)^{p/2} d\mu(t).$$

Now using the inequality

$$\left(\sum c_i^2\right)^{1/2} \geq \left(\sum_i^p c_i^p\right)^{1/p}, \quad c_i \geq 0, \quad p \geq 2$$

(which expresses the monotonicity of the l_p-norm) we obtain the needed estimate

$$\max_{\alpha_i=\pm 1} \left\| \sum_{i=1}^n \alpha_i x_i \right\| \geq \left(\sum_{i=1}^n \int_\Omega |x_i(t)|^p d\mu(t) \right)^{1/p} = \left(\sum_{i=1}^n \|x_i\|^p \right)^{1/p}.$$

This completes the proof of Orlicz's theorem. □

EXERCISE 4.2.1. We have proved that the spaces L_p have M-cotype $r = \max\{p, 2\}$. Show that if $s < r$, then L_p does not have M-cotype s.

EXERCISE 4.2.2. Show that the Dvoretzky-Rogers theorem is sharp, in the sense that in its statement the condition $\sum_{i=1}^\infty a_i^2 < \infty$ cannot be weakened.

EXERCISE 4.2.3. Clarify whether some analogue of Orlicz's theorem holds in the spaces $C[0,1]$ and L_∞.

EXERCISE 4.2.4. Use the Khinchin inequality to show that in any of the spaces $L_p[0,1]$, $1 \leq p < \infty$, the closure of the linear span of the sequence of Rademacher functions is a subspace isomorphic to l_2 (the Rademacher functions $r_n(t) = \text{sign} \sin(2^n \pi t)$ constitute one of the simplest models of a sequence of independent Bernoulli random variables).

EXERCISE 4.2.5. Prove that in the space $L_\infty[0,1]$ the preceding assertion is false. More precisely, in this space the closure of the linear span of the sequence of Rademacher functions $\{r_n\}$ is isomorphic to the space l_1.

§3. Absolutely Summing Operators. Grothendieck's Theorem

One of the possible ways of generalizing the results of the preceding section is to consider the convergence of a series in one topology, but compute the norms of its elements in another, weaker topology. This approach is quite productive and lies at the foundation of the deep and well developed theory of absolutely summing operators. We shall present here some ideas of this theory; for a more detailed exposition the reader should consult the monograph [70] and the references given therein. For applications to Banach spaces and function theory see [14], [52], [101].

THEOREM 4.3.1. *The canonical embedding operator $L_\infty[0,1] \to L_p[0,1]$ $(1 \leq p < \infty)$ maps unconditionally convergent series into p-absolutely convergent ones, (i.e., series with the property that $\sum_{i=1}^\infty \|x_i\|^p < \infty$).*

PROOF. To simplify the arguments we remark that the assertion of the theorem is equivalent to the following one: for any finite collection $\{x_i\}$ of elements of $L_\infty[0,1]$ one has the inequality

$$\max_{\alpha_i=\pm 1}\left\|\sum \alpha_i x_i\right\|_{L_\infty} \geq \gamma\left(\sum \|x_i\|_{L_p}^p\right)^{1/p}, \tag{1}$$

where $\gamma > 0$ does not depend on the choice of the elements x_i.

Thus, pick a finite set of elements $x_i(t)$ of $L_p[0,1]$. Using the fact that the norm of an element in l_p [resp. L_p] decreases [resp. increseas] when p is increased, we obtain the following chain of inequalities, starting from the left-hand side of (1):

$$\max_{\alpha_i=\pm 1}\left\|\sum \alpha_i x_i\right\|_{L_\infty} = \max_{\alpha_i=\pm 1}\sup_{t\in[0,1]}\left|\sum \alpha_i x_i(t)\right|$$

$$= \sup_{t\in[0,1]}\max_{\alpha_i=\pm 1}\left|\sum \alpha_i x_i(t)\right| = \sup_{t\in[0,1]}\sum |x_i(t)|$$

$$\geq \sup_{t\in[0,1]}\left(\sum |x_i(t)|^p\right)^{1/p} \geq \left(\int_0^1 \sum |x_i(t)|^p\,dt\right)^{1/p} = \left(\sum \|x_i\|_{L_p}^p\right)^{1/p}.$$

This proves (1), and hence the assertion of the theorem. □

EXERCISE 4.3.1. Show that in the preceding theorem the requirement of p-absolute convergence cannot be replaced by r-absolute convergence with $r < p$.

DEFINITION 4.3.1. A linear operator $T\colon X \to Y$ is said to be *absolutely summing* if it maps any unconditionally convergent series $\sum_{i=1}^\infty x_i$, $x_i \in X$, into an absolutely convergent series: $\sum_{i=1}^\infty \|Tx_i\| < \infty$.

As in many of the problems considered earlier in this book, the last definition can be translated into the language of finite collections of elements:

DEFINITION 4.3.2. A linear operator $T\colon X \to Y$ is said to be *absolutely summing* if there exists a constant K such that

$$\sum \|Tx_i\| \leq K \max_{\alpha_i=\pm 1}\left\|\sum \alpha_i x_i\right\| \tag{2}$$

for any finite collection $\{x_i\}$ of elements of X. The smallest value of the constant K is called the *absolutely summing norm* of the operator T, and is denoted by $\pi(T)$. If the operator T is not absolutely summing, we put $\pi(T) = \infty$. In other words,

$$\pi(T) = \sup\left\{\sum_{i=1}^n \|Tx_i\| \Big/ \max_{\alpha_i=\pm 1}\left\|\sum_{i=1}^n \alpha_i x_i\right\| : \{x_i\}_{i=1}^n \subset X,\ n \in \mathbf{N}\right\}. \tag{3}$$

EXERCISE 4.3.2. Show that the two preceding definitions give the same class of operators.

EXERCISE 4.3.3. Show that any absolutely summing operator is bounded and that $\|T\| \le \pi(T)$.

EXERCISE 4.3.4. Show that the set $\Pi(X,Y)$ of absolutely summing operators $T\colon X \to Y$ is a Banach space with respect to the norm $\pi(T)$ whenever the normed spaces X and Y are complete.

EXERCISE 4.3.5. Calculate the absolutely summing norm of the identity (embedding) operator $L_\infty \to L_1$.

EXERCISE 4.3.6. Show that the identity (embedding) operator $l_1 \to l_2$ is absolutely summing.

The fact observed in the last exercise is a simple particular consequence of the following more general assertion, which lies at the foundation of the theory of absolutely summing operators.

THEOREM 4.3.2 (GROTHENDIECK'S THEOREM). *Any bounded linear operator $T\colon l_1 \to l_2$ is absolutely summing; moreover,*

$$\pi(T) \le K\|T\|, \tag{4}$$

where K does not depend on T.

PROOF. By Definition 4.3.2, we need to show that

$$\sum \|Tx_i\| \le K\|T\| \cdot \max_{\alpha_i=\pm1} \left\|\sum \alpha_i x_i\right\|.$$

In the left-hand side of this inequality let us replace $\|Tx_i\|$ by the inner product (Tx_i, y_i), where $y_i \in l_2$, $\|y_i\| \le 1$:

$$\left|\sum (Tx_i, y_i)\right| \le K\|T\| \cdot \max_{\alpha_i=\pm1} \left\|\sum \alpha_i x_i\right\|.$$

Inequality (4) is equivalent to the last inequality being valid for all $x_i \in l_1$ and all $y_i \in l_2$, $\|y_i\| < 1$. Let us decompose the elements x_i, y_i and the operator T with respect to the standard bases $\{e_j\}$ and $\{\bar{e}_j\}$ in l_1 and l_2, respectively:

$$x_i = \sum_j x_{i,j} e_j, \quad Te_j = \sum_k t_{k,j} \bar{e}_k, \quad y_i = \sum_k y_{i,k} \bar{e}_k. \tag{5}$$

With no loss of generality we may assume that all the sums in (5) have only finitely many nonzero terms (thanks to the convergence of series, the passage to an infinitely number of elements can be accomplished by continuity). We shall assume also that $\|T\| \le 1$, i.e., $\sum_k t_{k,j}^2 \le 1$ for all $j \in \mathbf{N}$. Then, in the notations of (5), inequality (4) takes on the form

$$\left|\sum_{i,j,k} x_{i,j} t_{k,j} y_{i,k}\right| \le K \max_{\alpha_i=\pm1} \sum_j \left|\sum_i \alpha_i x_{i,j}\right| = K \max_{\alpha_i,\beta_j=\pm1} \left|\sum_{i,j} \alpha_i x_{i,j} \beta_j\right|, \tag{6}$$

where $\sum_k y_{i,k}^2 \le 1$ for all i and $\sum_k t_{k,j}^2 \le 1$ for all j.

Thus, all reduces to proving inequality (6). First, let us rename the quantities involved to arrive at the traditional form, namely, let us make the substitutions

$$x_{i,j} \to a_{i,j}, \quad y_{i,k} \to x_{i,k}, \quad t_{k,j} \to y_{j,k}, \quad \alpha_i \to s_i, \quad \beta_j \to t_j.$$

Then (6) becomes

$$\left|\sum_{i,j,k} a_{i,j} x_{i,k} y_{j,k}\right| \leq K \max_{s_i,t_j=\pm 1} \sum_{i,j} a_{i,j} s_i t_j,$$

where $\sum_k x_{i,k}^2 \leq 1$, $\sum_k y_{j,k}^2 \leq 1$. If we now put $x_i = \sum_k x_{i,k} e_k$ and $y_j = \sum_k y_{j,k} e_k$, where $\{e_k\}$ now denotes the standard orthonormal basis in l_2, we arrive at the following inequality, known as the *Grothendieck inequality*:

$$\left|\sum_{i,j} a_{i,j}(x_i, y_j)\right| \leq K \max_{s_i,t_j=\pm 1} \sum_{i,j} a_{i,j} s_i t_j, \tag{7}$$

where $(a_{i,j})$ is an arbitrary real $n \times n$ matrix (which can be assumed to be square by adding zeros for the missing entries) and where x_i and y_j are elements of the unit sphere of the space $l_2^{(n)}$.

Proof of Grothendieck's inequality. First of all let us remark that, thanks to the homogeneity of (7), it suffices to examine the case of matrices $(a_{i,j})$ for which

$$\max_{s_i,t_j=\pm 1} \sum_{i,j} a_{i,j} s_i t_j = 1,$$

or, equivalently,

$$-1 \leq \sum_{i,j} a_{i,j} s_i t_j \leq 1 \tag{8}$$

for all $s_i, t_j = \pm 1$. Consider the integral

$$\int_S \operatorname{sign}(x, u) \cdot \operatorname{sign}(y, u)\, dm(u), \tag{9}$$

where S is the unit sphere of $l_2^{(n)}$, m is the normalized Lebesgue measure on S, x, y are fixed elements of S, $u \in S$ is the integration variable, and $(\cdot,\cdot)$ denotes the inner product. Let us pass to spherical coordinates: if $u = \sum_{i=1}^n u_i e_i$, $u_i = (u, e_i)$, $\sum_{i=1}^n u_i^2 = 1$, then

$$\begin{aligned}
u_1 &= \sin\varphi_1 \cdot \sin\varphi_2 \cdots \sin\varphi_{n-2} \cdot \sin\varphi_{n-1}, \\
u_2 &= \cos\varphi_1 \cdot \sin\varphi_2 \cdots \sin\varphi_{n-2} \cdot \sin\varphi_{n-1}, \\
u_3 &= \cos\varphi_2 \cdots \sin\varphi_{n-2} \cdot \sin\varphi_{n-1}, \\
&\vdots \\
u_{n-1} &= \cos\varphi_{n-2} \cdot \sin\varphi_{n-1}, \\
u_n &= \cos\varphi_{n-1},
\end{aligned}$$

with $0 \leq \varphi_1 \leq 2\pi$, $0 \leq \varphi_i \leq \pi$, $i = 2, \ldots, n$. The Jacobian of this transformation of coordinates equals

$$J(\varphi) = \sin \varphi_2 \sin^2 \varphi_3 \cdot \ldots \cdot \sin^{n-2} \varphi_{n-1}.$$

With no loss of generality we may assume (thanks to the invariance of the Lebesgue measure under rotations of the sphere) that $x = e_1$, $y = \cos\theta\, e_1 + \sin\theta\, e_2$, where θ is the angle made by the vectors x and y. After the change of variables the integral (9) becomes

$$\frac{1}{C} \int_0^{2\pi} \int_0^{\pi} \cdots \int_0^{\pi} \operatorname{sign} u_1 \cdot \operatorname{sign}(u_1 \cos\theta + u_2 \sin\theta) J(\varphi)\, d\varphi_1 \ldots d\varphi_{n-1},$$

where

$$C = \int_0^{2\pi} \int_0^{\pi} \cdots \int_0^{\pi} J(\varphi)\, d\varphi_1 \ldots d\varphi_{n-1}.$$

After the usual simplifications (in particular, cancellation of all factors that do not contain φ_1), we find that

$$\frac{1}{2\pi} \int_0^{2\pi} \operatorname{sign}\left[\sin\varphi_1 \cdot \sin(\varphi_1 + \theta)\right] d\varphi_1 = 1 - \frac{2\theta}{\pi} = \frac{2}{\pi} \arcsin(x, y),$$

i.e.,

$$\int_S \operatorname{sign}(x, u) \cdot \operatorname{sign}(y, u)\, dm(u) = \frac{2}{\pi} \arcsin(x, y). \tag{10}$$

Now, armed with the auxiliary formula (10), let us return to Grothendieck's inequality. Since the maximum of the left-hand side of (7), taken for fixed $a_{i,j}$ over $x_i, y_j \in B(l_2^{(n)})$, can be attained only for $\|x_i\| = \|y_j\| = 1$, we shall assume that $x_i, y_j \in S(= S(l_2^{(n)}))$. Given an arbitrary number $m \in \mathbf{N}$, let us go to inequality (8), which is supposed to hold for all $s_i, t_j = \pm 1$, substitute in it

$$s_i = \prod_{k=1}^{m} \operatorname{sign}(x_i, u^{(k)}), \quad t_j = \prod_{k=1}^{m} \operatorname{sign}(y_j, u^{(k)}),$$

and then integrate with respect to the variables $u^{(k)} \in S$, $k = 1, \ldots, m$. Using (10), we see that, for any $m \in \mathbf{N}$,

$$-1 \leq \sum_{i,j} a_{i,j} \left[\frac{2}{\pi} \arcsin(x_i, y_j)\right]^m \leq 1. \tag{11}$$

Now let us multiply each inequality (11) with m odd by $(\pi/2)^m(-1)^{m-1}/m!$, take into account the sign of the factor, and add the resulting two-sided inequalities. We obtain

$$\left| \sum_{i,j} \sin(\arcsin(x_i, y_j)) \cdot a_{i,j} \right| \leq \sinh \frac{\pi}{2},$$

or, finally,

$$\left|\sum_{i,j} a_{i,j}(x_i, y_j)\right| \leq \sinh\frac{\pi}{2} \cdot \max_{s_i,t_j=\pm 1} \sum_{i,j} a_{i,j} s_i t_j.$$

Thus, we have established Grothendieck's inequality with the constant $K = \sinh\frac{\pi}{2}$. Since Grothendieck's inequality is equivalent to inequality (4), this completes the proof of the theorem. □

The exact value of the constant K (*Grothendieck's constant* K_{G}) is not known. It is only known that (see [70])

$$\frac{\pi}{2} \leq K_{\mathrm{G}} \leq \frac{\pi}{2\ln(\sqrt{2}+1)}.$$

EXERCISE 4.3.7. Let $A\colon l_1^{(n)} \to l_2^{(n)}$, $B\colon l_\infty^{(n)} \to l_1^{(n)}$, and $C\colon l_2^{(n)} \to l_\infty^{(n)}$ be arbitrary linear operators. Prove the following equivalent form of Grothendieck's inequality:

$$\operatorname{trace}(ABC) \leq K_{\mathrm{G}}\|A\| \cdot \|B\| \cdot \|C\|.$$

DEFINITION 4.3.3. A locally convex topological linear space E is said to be *nuclear* if for any continuous seminorm p on E there exists a continuous seminorm $q > p$ on E for which the identity operator $I\colon (E, q) \to (E, p)$ is absolutely summing.

EXERCISE 4.3.8. Show that the space of functions analytic in an open domain, equipped with the topology of uniform convergence on compact sets, is nuclear.

EXERCISE 4.3.9. Show that in a nuclear space any unconditionally convergent series $\sum_{k=1}^{\infty} x_k$ converges absolutely, i.e., $\sum_{k=1}^{\infty} p(x_k) < \infty$ for any continuous seminorm p.

CHAPTER 5
ORLICZ'S THEOREM AND THE STRUCTURE OF FINITE-DIMENSIONAL SUBSPACES

In the preceding chapters we have often encountered assertions of the following type: from a certain property of the sum of a finite number of elements follows a certain theorem on series in an infinite-dimensional space. Since any finite set of elements lies in some finite-dimensional subspace, in all such assertions only special features of the structure of the finite-dimensional subspaces of the ambient infinite-dimensional space considered play a role. In the first section of this chapter we will present a modern convenient language for describing what kind of finite-dimensional subspaces a given infinite-dimensional space has. In the sequel we shall actively employ this language and, in particular, give a complete description of the spaces in which an analogue of Orlicz's theorem on unconditionally convergent series holds.

§1. Finite Representability

DEFINITION 5.1.1. Let X and Y be Banach spaces. The *Banach-Mazur distance* between X and Y is the number $d(X,Y) = \inf_T \{\|T\|, \|T^{-1}\|\}$, where the infimum is taken over all isomorphisms $T\colon X \to Y$. If X and Y are not isomorphic, we put $d(X,Y) = +\infty$.

Let us list the simplest properties of the Banach-Mazur distance.

1) $d(X,Y) \geq 1$ for any X, Y. If X and Y are isometrically isomorphic, then $d(X,Y) = 1$.

2) The Banach-Mazur distance is symmetric: $d(X,Y) = d(Y,X)$.

3) The multiplicative triangle inequality holds:

$$d(X,Y) \cdot d(Y,Z) \geq d(X,Z).$$

The last property follows from the definition of $d(\cdot,\cdot)$ and the inequality $\|BA\| \leq \|B\| \cdot \|A\|$, which holds for all bounded operators $A\colon X \to Y$ and $B\colon Y \to Z$.

DEFINITION 5.1.2. A Banach space X is said to be *finitely representable in a space* Y (and one writes $X \xrightarrow{f} Y$) if for any $\varepsilon > 0$ and any finite-dimensional subspace $Z \subset X$ there exists a finite-dimensional subspace $Z_1 \subset Y$ such that $d(Z, Z_1) < 1 + \varepsilon$.

Notice that if X admits an isometric embedding in Y, then $X \xrightarrow{\mathrm{f}} Y$. Hence, for any $p \geq 1$ the space l_p is finitely representable in $L_p[0,1]$. However, finite reprensentability is not equivalent to the existence of an isometric embedding.

PROPOSITION 5.1.1. *$L_p[0,1] \xrightarrow{\mathrm{f}} l_p$ for any $p \geq 1$.*

PROOF. Let $Z \subset L_p[0,1]$ be a subspace, $\dim Z = n$. Pick a basis $\{z_k\}_{k=1}^n$ in Z. Since all the norms on a finite-dimensional space are equivalent to one another, there exists a constant C such that

$$\left\|\sum_{k=1}^n t_k z_k\right\| \geq C \sum_{k=1}^n |t_k|$$

for all collections of numbers $\{t_k\}_{k=1}^n$.

A measurable function will be said to be *simple* if it takes only a finite number of values. Take an arbitrary $\varepsilon > 0$ and choose a $\delta > 0$ such that $(1+\delta)/(1-\delta) < 1+\varepsilon$. Approximate each element z_k in the L_p-metric by a simple function y_k such that $\|z_k - y_k\| \leq C\delta$. Set $Y = \mathrm{Lin}\,\{y_1, \dots, y_n\}$. Let us show that $d(Z,Y) \leq 1+\varepsilon$. To this end let us introduce the operator $T\colon Z \to X$ by the rule

$$T\left(\sum_{k=1}^n t_k z_k\right) = \sum_{k=1}^n t_k y_k.$$

Then

$$\left\|T\left(\sum_{k=1}^n t_k z_k\right)\right\| = \left\|\sum_{k=1}^n t_k y_k\right\| \leq \left\|\sum_{k=1}^n t_k z_k\right\| + \sum_{k=1}^n |t_k|\,\|y_k - z_k\|$$
$$\leq \left\|\sum_{k=1}^n t_k z_k\right\| + C\delta \sum_{k=1}^n |t_k| \leq (1+\delta)\left\|\sum_{k=1}^n t_k z_k\right\|,$$

which means that $\|T\| \leq 1+\delta$. On the other hand,

$$\left\|T\left(\sum_{k=1}^n t_k z_k\right)\right\| = \left\|\sum_{k=1}^n t_k y_k\right\| \geq \left\|\sum_{k=1}^n t_k y_k\right\| - C\delta \sum_{k=1}^n |t_k| \geq (1-\delta)\left\|\sum_{k=1}^n t_k z_k\right\|.$$

Hence, T is an isomorphism, $\|T\| \leq 1+\delta$, and $\|T^{-1}\| \leq 1/(1-\delta)$. Consequently,

$$d(Z,Y) \leq \|T\| \cdot \|T^{-1}\| \leq \frac{1+\delta}{1-\delta} < 1+\varepsilon.$$

It remains to show that Y embeds isometrically in l_p. Since the functions y_k are simple, we can find disjoint measurable sets A_j, $j = 1, \dots, N$, such that $\bigcup_{j=1}^N A_k = [0,1]$ and $\mu(A_j) > 0$ for $j = 1, \dots, N$, and on each A_j each of the functions y_k is constant. Let χ_{A_j} denote the indicator function of the set A_j. Then $\mathrm{Lin}\{\chi_{A_1}, \dots, \chi_{A_N}\}$ is a subspace of $L_p[0,1]$ that is isometric to $l_p^{(N)}$. Since $Y \subset \mathrm{Lin}\,\{\chi_{A_1}, \dots, \chi_{A_N}\}$, we conclude that Y is isometrically ismorphic to a subspace of l_p, which completes the proof of the proposition. □

PROPOSITION 5.1.2. *The space $C[0,1]$ is finitely representable in the space c_0.*

PROOF. On proceeds as in the proof of Proposition 4.1.1, but use piecewise-linear functions instead of simple ones. □

From the last proposition and the theorem on the universality of the space $C[0,1]$ (see [55]) we obtain

COROLLARY 5.1.1. *Any Banach space is finitely representable in the space c_0.*

EXERCISE 5.1.1. Show that if $X \xrightarrow{f} Y$ and $Y \xrightarrow{f} Z$, then $X \xrightarrow{f} Z$.

EXERCISE 5.1.2. Give an example of spaces X and Y such that $X \xrightarrow{f} Y$, but it is not true that $Y \xrightarrow{f} X$.

EXERCISE 5.1.3. Show that if $X \xrightarrow{f} l_2$, then X is isometrically ismorphic to a Hilbert space.

EXERCISE 5.1.4. Give an example of spaces X and Y that are not isometrically isomorphic, and such that $d(X,Y) = 1$.

EXERCISE 5.1.5. Show that if $X \xrightarrow{f} L_p[0,1]$, then the space X embeds isometrically in $L_p[0,1]$.

EXERCISE 5.1.6. Show that if X is a finite-dimensional space and the unit sphere $S(X)$ is a polyhedron, then X embeds isometrically in c_0.

EXERCISE 5.1.7. Show that any two-dimensional normed space is finitely representable in l_1.

EXERCISE 5.1.8. Without relying on Exercise 5.1.5, show that Orlicz's theorem on unconditionally convergent series in L_p remains valid for any space that is finitely representable in L_p

EXERCISE 5.1.9. Calculate $d(l_2^{(n)}, l_p^{(n)})$.

EXERCISE 5.1.10. Show that the set of all n-dimensional normed spaces, equipped with the metric $\log d(X,Y)$, is a compact space (here isometrically isomorphic spaces are identified).

EXERCISE 5.1.11. Show that for any nonseparable space X there exists a separable space Y such that $X \xrightarrow{f} Y$ and $Y \xrightarrow{f} X$.

EXERCISE 5.1.12. Prove the *principle of local reflexivity*, i.e., that $X^{**} \xrightarrow{f} X$ for any space X.

§2. The Space c_0, C-Convexity, and Orlicz's Theorem

Let us consider the Banach space c_0, that is, the space of all sequence $x = (x_1, x_2, \dots)$, $x_i \in \mathbf{R}$, such that $x_i \to 0$, equipped with the norm $\|x\| = \max_j |x_j|$. Let $\{e_k\}_{k=1}^{\infty}$ be the standard basis of c_0. Clearly, if a_k are positive numbers such that $a_k \to 0$ when $k \to \infty$ and if $\alpha_k = \pm 1$, then

$$\left\| \sum_{k=n}^{n+m} \alpha_k a_k e_k \right\| = \max_{n \le k \le n+m} a_k \to 0$$

when $n \to \infty$. Hence, the series $\sum_{k=1}^{\infty} a_k e_k$ converges unconditionally for any sequence of numbers $\{a_k\}_{k=1}^{\infty}$ such that $a_k \to 0$. We see that in c_0 the unconditional convergence of a series $\sum_{k=1}^{\infty} y_k$ imposes no constraints on the norms of its terms, except for the trivial condition $\lim_{k\to\infty} \|y_k\| = 0$. It follows that no analogue of Orlicz's theorem holds in c_0. We shall demonstrate now that this is the case not only for the space c_0, but also for any space in which c_0 is finitely representable.

THEOREM 5.2.1. *Let X be a Banach space such that $c_0 \xrightarrow{\text{f}} X$. Let $\{a_k\}_{k=1}^{\infty}$ be a sequence of positive numbers that converges to zero. Then in X there exists an unconditionally convergent series $\sum_{k=1}^{\infty} x_k$ such that $\|x_k\| = a_k$ for all k.*

PROOF. Choose an increasing sequence of natural numbers n_k such that $a_i < 2^{-k}$ for all $i > n_k$. For such a choice the series

$$\sum_{k=1}^{\infty} \max_{n_n < i \le n_k+1} a_i \tag{1}$$

converges. Since $l_{\infty}^{(n)}$ is a finite-dimensional subspace of c_0 and c_0 is finitely representable in X, it follows that one can find subspaces in X that are almost isometric to the spaces

$$l_{\infty}^{(n_1)},\ l_{\infty}^{(n_2-n_1)},\ l_{\infty}^{(n_3-n_2)},\ \dots .$$

In other words, one can exhibit a sequence of vectors $\{e_i\}_{i=1}^{\infty}$ in X such that the inequalities

$$\max_{n_n < i \le n_k+1} |t_i| \le \left\| \sum_{i=n_k+1}^{n_{k+1}} t_i e_i \right\| \le 2 \max_{n_n < i \le n_k+1} |t_i|$$

will hold for any $i \in \mathbf{N}$ and any collection of numbers t_i. This implies, in particular, that $1 \le \|e_i\| \le 2$. Now we define the sought-for series $\sum_{k=1}^{\infty} x_k$ by the rule $x_k = a_k e_k / \|e_k\|$, $k = 1, 2, \dots$. Obviously, $\|x_k\| = a_k$, and so it remains to verify that our series converges unconditionally. Let $\alpha_k = \pm 1$ be an arbitrary collection of signs and let n and m be natural numbers, with $n > n_k$. Then

$$\left\| \sum_{i=n}^{n+m} \alpha_i x_i \right\| = \left\| \sum_{i=n}^{n+m} \alpha_i a_i \frac{e_i}{\|e_i\|} \right\| \le \sum_{j=k}^{\infty} 2 \max_{n_j < i \le n_{j+1}} \frac{a_i}{\|e_i\|} \le 2 \sum_{j=k}^{\infty} \max_{n_j < i \le n_{j+1}} a_i .$$

Thanks to the convergence of the series (1), we have

$$\lim_{n,m\to\infty}\left\|\sum_{i=n}^{n+m}\alpha_i x_i\right\| = 0,$$

which by the Cauchy criterion means that the series $\sum_{i=1}^{\infty}\alpha_i x_i$ converges. □

Thus, we have exhibited a class of spaces in which for sure no analogue of Orlicz's theorem is valid, namely, the class of spaces in which c_0 is finitely representable. It is natural to look now at the opposite class.

DEFINITION 5.2.1. A space X is said to be *C-convex* if c_0 is not finitely representable in X.

The next simple lemma, which is a reformulation of the definition of C-convexity, is given without proof.

LEMMA 5.2.1. *For the space X to be C-convex it is necessary and sufficient that there exist a natural number $n > 0$ and a number $\varepsilon > 0$ such that*

$$d\left(A, l_\infty^{(n)}\right) \geq 1+\varepsilon$$

for any n-dimensional subspace $A \subset X$. □

LEMMA 5.2.2. *Let X be a Banach space. The following two conditions are equivalent:*

(a) *There exist $n \in \mathbf{N}$ and $\varepsilon > 0$ such that*

$$d\left(A, l_\infty^{(n)}\right) \geq 1+\varepsilon$$

for any n-dimensional subspace $A \subset X$.

(b) *There exist $n \in \mathbf{N}$ and $\varepsilon_1 > 0$ such that, for any collection of elements $\{x_i\}_{i=1}^n$, $\|x_i\| \geq 1$, one can find signs $\nu_k = \pm 1$ for which*

$$\left\|\sum_{i=1}^{n}\nu_i x_i\right\| \geq 1+\varepsilon_1.$$

PROOF. First let us show that (b) ⇒ (a). Let n and ε_1 be as in condition (b) and A be an n-dimensional subspace of X. Further, let $T\colon l_\infty^{(n)} \to A$ be an arbitrary isomorphism between $l_\infty^{(n)}$ and A. Denote by e_i the vectors of the standard basis of $l_\infty^{(n)}$. Then

$$\|Te_i\| \geq \frac{1}{\|T^{-1}\|}, \quad i = 1,\ldots,n.$$

Choose signs $\nu_i = \pm 1$ such that

$$\left\| \sum_{i=1}^{n} \nu_i T e_i \right\| \geq \frac{1+\varepsilon_1}{\|T^{-1}\|}.$$

Thanks to the relation $\| \sum_{i=1}^{n} \nu_i e_i \|_\infty = 1$, this yields

$$\|T\| \geq \frac{\|T(\sum_{i=1}^{n} \nu_i e_i)\|_X}{\| \sum_{i=1}^{n} \nu_i e_i \|_\infty} = \left\| \sum_{i=1}^{n} \nu_i T e_i \right\|_X \geq \frac{1+\varepsilon_1}{\|T^{-1}\|}.$$

Therefore, $\|T\| \cdot \|T^{-1}\| \geq 1+\varepsilon_1$, which in view of the arbitrariness of the operator T means that $d\left(l_\infty^{(n)}, A\right) \geq 1+\varepsilon_1$ for any n-dimensional subspace $A \subset X$.

(a) $\Rightarrow$ (b). We shall argue by reductio ad absurdum. Suppose that for any $n \in \mathbf{N}$ and any $\varepsilon > 0$ there exists a collection of elements $\{x_i\}_{i=1}^n$, $\|x_i\| \geq 1$, $i = 1, \ldots, n$, such that the inequality

$$\left\| \sum_{i=1}^{n} \nu_i x_i \right\| < 1 + \varepsilon$$

holds for any choice of signs $\nu_i = \pm 1$. Consider the operator $T\colon l_\infty^{(n)} \to \operatorname{Lin}\{x_i\}_{i=1}^n$, defined by $Te_i = x_i$, $i = 1, \ldots, n$. We claim that $\|T\| \leq 1+\varepsilon$. Indeed, let x be an element of the unit ball of $l_\infty^{(n)}$. Write x as a convex combination of elements of the form $\sum_{i=1}^n \nu_i e_i$, $\nu_i = \pm 1$:

$$x = \sum_{\nu_i = \pm 1} \left(\lambda_{\nu_1, \ldots, \nu_n} \sum_{i=1}^{n} \nu_i e_i \right),$$

where

$$\lambda_{\nu_1, \ldots, \nu_n} \geq 0 \quad \text{and} \quad \sum_{\nu_i = \pm 1} \lambda_{\nu_1, \ldots, \nu_n} = 1.$$

The existence of such a representation is guaranteed by the fact that the unit ball of $l_\infty^{(n)}$ is the polyhedron (more precisely, the cube) with the vertices $\sum_{i=1}^n \nu_i e_i$. We have

$$\|Tx\| = \left\| \sum_{\nu_i = \pm 1} \left(\lambda_{\nu_1, \ldots, \nu_n} \sum_{i=1}^{n} \nu_i T e_i \right) \right\| \leq \sum_{\nu_i = \pm 1} \left(\lambda_{\nu_1, \ldots, \nu_n} \left\| \sum_{i=1}^{n} \nu_i e_i \right\| \right) \leq 1 + \varepsilon.$$

Since $x \in B(l_\infty^{(n)})$ is arbitrary, we conclude that $\|T\| \leq 1+\varepsilon$. Now let us show that $\|T^{-1}\| \leq (1-\varepsilon)^{-1}$. Let $x = \sum_{i=1}^n a_i x_i$ and set $a_{i_0} = \max_i |a_i|$. Consider the auxiliary vector $y = \sum_{i=1}^n b_i x_i$, where $b_i = a_i$ if $i \neq i_0$ and $b_{i_0} = -a_{i_0}$. Then

$$\frac{\|x\| + \|y\|}{2} \geq \left\| \frac{x-y}{2} \right\| = \|a_{i_0} x_{i_0}\| \geq |a_{i_0}| = \max_i |a_i| = \left\| \sum_{i=1}^{n} a_i e_i \right\|.$$

Consequently, $\|x\| \geq 2\|\sum_{i=1}^n a_i e_i\| - \|y\|$, and since

$$\|y\| = \left\|\sum_{i=1}^n b_i x_i\right\| = \left\|T\left(\sum_{i=1}^n b_i e_i\right)\right\| \leq (1+\varepsilon)\left\|\sum_{i=1}^n b_i e_i\right\|$$
$$= (1+\varepsilon)\max_i |a_i| = (1+\varepsilon)\left\|\sum_{i=1}^n a_i e_i\right\|,$$

we conclude that $\|x\| \geq (1-\varepsilon)\|\sum_{i=1}^n a_i e_i\| = (1-\varepsilon)\|T^{-1}(x)\|$, whence $\|T^{-1}\| \leq (1-\varepsilon)^{-1}$. Therefore,

$$d\left(l_\infty^{(n)}, \operatorname{Lin}\{x_i\}_{i=1}^n\right) \leq \|T\| \cdot \|T^{-1}\| \leq \frac{1+\varepsilon}{1-\varepsilon},$$

contrary to (a). □

DEFINITION 5.2.2. The *measure of C-convexity* of the space X is the function $C(n, X) : \mathbf{N} \to \mathbf{R}_+$ defined by the formula

$$C(n, X) = \inf\left\{\max\left\{\left\|\sum_{k=1}^n \alpha_k x_k\right\| : \alpha_k = \pm 1\right\} : \|x_i\| \geq 1\right\}.$$

It is readily seen that $C(n, X) \geq 1$ for all n. From the two preceding lemmas it follows that for a space X to be C-convex it is necessary and sufficient that the strict inequality $C(n, X) > 1$ holds for some $n = n(X) \in \mathbf{N}$.

THEOREM 5.2.2. *The sequence $C(n, X)$ is nondecreasing; the numbers $C(n, X)$ form a "semimultiplicative sequence," i.e.,*

$$C(n \cdot m, X) \geq C(n, X) \cdot C(m, X).$$

PROOF. The first assertion is an immediate consequence of the inequality $\max\{\|x+y\|, \|x-y\|\} \geq \|x\|$. Let us prove the semimultiplicativity property. Let $\{x_i\}_{i=1}^{nm}$ be elements of X, $\|x_i\| \geq 1$. Write the set $\{x_i\}_{i=1}^{nm}$ as a rectangular table: $y_{r,l}$, where r runs from 1 to n and l runs from 1 to m. For each l choose numbers $\nu_{r,l} = \pm 1$, $r = 1, \ldots, n$, such that

$$\left\|\sum_{r=1}^n \nu_{r,l} y_{r,l}\right\| \geq C(n, X),$$

and denote the sum $\sum_{r=1}^n \nu_{r,l} y_{r,l}$ by z_l. Then

$$\max_{\theta_i = \pm 1}\left\|\sum_{i=1}^{nm} \theta_i x_i\right\| \geq \max_{\alpha_l = \pm 1}\left\|\sum_{l=1}^m \alpha_l z_l\right\| \geq C(m, X) \cdot \min_l \|z_l\| \geq C(m, X) \cdot C(n, X),$$

which in view of the arbitrariness of the elements x_k completes the proof of semimultiplicativity. □

COROLLARY 5.2.1. *If X is a C-convex space, then $\lim_{n\to\infty} C(n, X) = \infty$.*

COROLLARY 5.2.2. *Let X and Y be isomorphic Banach spaces. Then from the C-convexity of one of them follows the C-convexity of the other (stability of the class of C-convex spaces under isomorphisms).*

Recall that a Banach space is said to have *M-cotype p with constant $C > 0$* if the inequality

$$\max_{\nu_i=\pm 1} \left\| \sum_{i=1}^n \nu_i x_i \right\| \geq C \left(\sum_{i=1}^n \|x_i\|^p \right)^{1/p}$$

holds for any finite collection of elements $\{x_i\}_{i=1}^n$.

EXERCISE 5.2.1. Show that if a space X has M-cotype, then X is C-convex; that if $p_2 > p_1$ and X has M-cotype p_1, then it also has cotype p_2; that the M-cotype of any space cannot be smaller than 1; and that the M-cotype of a space X is inherited by all spaces that are finitely representable in X.

In the Chapter 4 we have shown (Theorem 4.2.2) that a space has M-cotype p if and only if Orlicz's theorem holds with exponent p in that space. Hence, in order to prove Orlicz's theorem for C-convex spaces is suffices to show that these spaces have an M-cotype.

THEOREM 5.2.3. *Let X be a C-convex space. Then there exist numbers $p \in \mathbf{N}$ and $C > 0$ such that X has M-cotype p with constant C.*

PROOF. Choose a natural number $n_1 > 1$ such that $C(n_1, X) = 1 + \delta > 1$. Set $n_0 = 1$, $n_k = (n_1)^k$, $k = 1, 2, 3, \ldots$. By Theorem 5.3.1, $C(n_k, X) \geq (1+\delta)^k$. Pick a natural number p such that $(1+\delta)^{p-1} \geq n_1$. Let us show that X has M-cotype p with the constant $C = 1/n_1$.

Take an arbitary collection $\{x_i\}_{i=1}^n$ of elements of X. Partition the index set $\{1, \ldots, n\}$ into disjoint sets A_k, $k = 0, 1, 2, \ldots$ as follows:

$$A_k \stackrel{\text{def}}{=} \left\{ j \colon \frac{\left(\sum_{i=1}^n \|x_i\|^p\right)^{1/p}}{n_k} \geq \|x_j\| > \frac{\left(\sum_{i=1}^n \|x_i\|^p\right)^{1/p}}{n_{k+1}} \right\}.$$

Denote by m_k the number of elements in A_k. Then

$$\sum_{j \in A_k} \|x_j\|^p \leq m_k \frac{\sum_{i=1}^n \|x_i\|^p}{(n_k)^p},$$

whence

$$\sum_{j=1}^n \|x_j\|^p = \sum_{k=0}^\infty \sum_{j \in A_k} \|x_j\|^p \leq \sum_{k=0}^\infty m_k \frac{\sum_{i=1}^n \|x_i\|^p}{(n_k)^p}.$$

It follows that $\sum_{k=0}^\infty m_k/(n_k)^p \geq 1$, or, equivalently, $\sum_{k=0}^\infty m_k/n_{kp} \geq 1$. This shows that there exists a number $k = k_0$ for which $m_{k_0} \geq n_{k_0(p-1)}$. (Indeed, if $m_k < n_{k(p-1)}$ for all k, then $m_0 = 0$, $m_k/n_{kp} < 1/(n_1)^k$, and $\sum_{k=0}^\infty m_k/n_{kp} < 1$.)

Recall that

$$\max_{\alpha_i=\pm 1}\left\|\sum_{i=1}^{n}\alpha_i x_i\right\| \geq \max_{\alpha_i=\pm 1}\left\|\sum_{i\in A}\alpha_i x_i\right\|$$

for any set $A \subset \{1,\ldots,n\}$. We are now ready to establish the inequality appearing in the definition of the M-cotype:

$$\max_{\alpha_i=\pm 1}\left\|\sum_{i=1}^{n}\alpha_i x_i\right\| \geq \max_{\alpha_j=\pm 1}\left\|\sum_{j\in A_k}\alpha_j x_j\right\| \geq \min_{j\in A_k}\|x_j\|\cdot C(m_{k_0},X)$$

$$> \frac{\left(\sum_{i=1}^{n}\|x_i\|^p\right)^{1/p}}{n_{k_0+1}}\cdot C(n_{k_0(p-1)},X) \geq \frac{\left(\sum_{i=1}^{n}\|x_i\|^p\right)^{1/p}}{n_{k_0+1}}\cdot(1+\delta)^{k_0(p-1)}$$

$$= \left(\frac{(1+\delta)^{p-1}}{n_1}\right)^{k_0}\cdot\frac{1}{n_1}\left(\sum_{i=1}^{n}\|x_i\|^p\right)^{1/p} \geq \frac{1}{n_1}\left(\sum_{i=1}^{n}\|x_i\|^p\right)^{1/p}.$$

Thus, X has M-cotype p with constant $1/n_1$, as claimed. □

Combining this result with Theorem 5.2.1 we obtain

COROLLARY 5.2.3. *If in a Banach space unconditional convergence imposes some constraint on the character of the convergence to zero of the general term of the series, then in that space unconditional convergence imposes a constraint of p-absolute convergence type.*

EXERCISE 5.2.2. Show that if for some n one has the inequality $C(n,X) \geq n^{1/p}$, then X has M-cotype $p+\varepsilon$ for any $\varepsilon > 0$.

EXERCISE 5.2.3. Provide an example of a space which has M-cotype $2+\varepsilon$ for any $\varepsilon > 0$, but does not have M-cotype 2.

The results on C-convexity given in this section belong to B. Maurey and G. Pisier [57]. The connection between these notions and the theory of series was also studied by S. A. Rakov [74], [75].

§3. Survey of Results on Type and Cotype

We have shown that in a Banach space the validity of an analogue of Orlicz's theorem on unconditional convergence of series is equivalent to C-convexity as well as to the existence of an M-cotype. Chobanyan's result (Theorem 3.4.2) and the results of Chapter 7 exhibit relationships between conditional convergence and concepts similar to those of C-convexity and M-cotype, namely, B-convexity and infratype.

The theory of type and cotype is one of the modern directions in Banach space theory, direction that is closely connected with interesting domains of functional analysis such as the theory of superproperties, the geometry of the unit sphere of a Banach space, or the theory of Banach-space valued stochastic processes. For this reason, the present section may be regarded as a link between the theory of series and many other actual problems of Banach space theory. To keep within the framework of our monograph, we wrote this section as a brief survey. Although sketches of proofs will be provided for some of the results, the bulk of the results is given without proofs. Many of the results listed below can be found in [99] or [101].

DEFINITION 5.3.1. A space X is said to have *cotype p with constant* $C > 0$ if the inequality

$$\mathbf{E}\left\|\sum_{i=1}^{n} r_i x_i\right\| \geq C\left(\sum_{i=1}^{n} \|x_i\|^p\right)^{1/p}$$

holds for any finite collection of elements $\{x_i\}_{i=1}^n \subset X$.

It is readily seen that if the space X has cotype p, then it has M-cotype p. It is not know whether for a infinite-dimensional spaces the existence of an M-cotype p is equivalent to the existence of a cotype p. The only known result in this direction is the following assertion.

THEOREM 5.3.1 [57]. *Suppose the space X has M-cotype p. Then X has cotype p_1 for any $p_1 > p$.* □

Recall that a space X is said to have *type p with constant C* if the inequality $\mathbf{E}\|\sum_{i=1}^n r_i x_i\| \leq C\left(\sum_{i=1}^n \|x_i\|^p\right)^{1/p}$ holds for any collection of elements $\{x_i\}_{i=1}^n \subset X$.

DEFINITION 5.3.2. A space X is said to have *infratype* $p > 1$ *with constant* C if the inequality

$$\min_{\alpha_i \pm 1}\left\|\sum_{i=1}^{n} \alpha_i x_i\right\| \leq C\left(\sum_{i=1}^{n} \|x_i\|^p\right)^{1/p}$$

holds for any finite collection of elements $\{x_i\}_{i=1}^n \subset X$.

The type and the infratype are related in the same way as the cotype and the M-cotype:

THEOREM 5.3.2. *If a space has type p, then it also has infratype p. If a space has infratype p, then it has type $p - \varepsilon$ for any $\varepsilon > 0$* (see [57]). □

Recently M. Talagrand proved the following result [89]: *for $p < 2$, a space has infratype p if and only if it has type p.*

Let us list some elementary properties of the type and cotype. The type, infratype, cotype, and M-cotype of a space X are inherited by all it subspaces as well as by all spaces that are finitely representable in X. The type and infratype of a space are inherited by quotient spaces, whereas the cotype and M-cotype are not.

Finite-dimensional spaces have M-cotype 1, cotype 2, type 2, and any infratype. The spaces L_p with $1 < p < \infty$ have type = cotype = $\min\{2, p\}$. The cotype and M cotype of the spaces L_p with $1 \leq p < \infty$ coincide and are equal to $\max\{2, p\}$. Dvoretzky's theorem (see Chapter 6) has the following simple consequence:

THEOREM 5.3.3. *Let X be an infinite-dimensional Banach space. If X has infratype p, then $p \leq 2$, and if X has M-cotype q, then $q \geq 2$.* □

Hilbert spaces can be characterized in term of type and cotype.

THEOREM 5.3.4 [49]. *An infinite-dimensional Banach space has simultaneously type 2 and cotype 2 if and only if it is isomorphic to a Hilbert space.* □

THEOREM 5.3.5 [57]. *Let X be an infinite-dimensional Banach space and let p_X denote the supremum of the types of X. Then for any $p \in [p_X, 2]$ the space l_p is finitely representable in X.* □

DEFINITION 5.3.3. The space X is said to be *B-convex* if the space l_1 is not finitely representable in X.

If the space X is B-convex, then so are its dual X^*, all spaces that are finitely representable in X, and all subspaces as well as all quotient spaces of X.

The next result can be established in much the same way as the corresponding assertion concerning C-convexity.

THEOREM 5.3.6. *A space X is B-convex if and only if there exist $n \in \mathbf{N}$ and $\varepsilon > 0$ such that the inequality $\min_{\nu_i=\pm 1} \|\sum_{i=1}^n \nu_i x_i\| \leq (1-\varepsilon)n$ holds for any collection of elements $\{x_i\}_{i=1}^n \subset B(X)$.* □

COROLLARY. *If a space has infratype $p > 1$, then it is B-convex.* □

As measures of B-convexity of a space X we will take the quantities

$$b(n, X) = \sup\left\{\min_{\nu_i=\pm 1}\left\|\sum_{i=1}^n \nu_i x_i\right\| : \{x_i\}_{i=1}^n \subset B(X)\right\}.$$

From Theorem 5.3.6 it follows that for the space X to be B-convex it is necessary and sufficient that there exist $n \in \mathbf{N}$ such that $b(n, X) < n$.

THEOREM 5.3.7. [71]. *Let X be a B-convex space. Then X has an infratype $p > 1$.* □

To prove this assertion one needs first to establish the inequality

$$b(n \cdot m, X) \leq b(n, X) \cdot b(m, X),$$

and then follow the scheme of proof of Theorem 5.2.3. □

Between type and cotype there is a certain duality (see [53, page 79]):

THEOREM 5.3.8. *Let X have type $p > 1$. Then X^* has cotype q, where $\frac{1}{p} + \frac{1}{q} = 1$.* □

This duality is not complete: the space L_1 has cotype 2, but $(L_1)^* = L_\infty$ is not B-convex.

The concept of B-convexity was introduced by A. Beck in 1962 [7] and was subsequently investigated by many authors (see [25], [26]). For a long time it was not known whether a nonreflexive space can be B-convex. The first such example was constructed by R. C. James in 1974 [29].

EXERCISE 5.3.1. Prove theorems 5.3.3 and 5.3.6–5.3.8.

EXERCISE 5.3.2. Show that any B-convex space is C-convex.

EXERCISE 5.3.3. Show that the C-convex space l_1 has c_0 among its quotient spaces.

EXERCISE 5.3.4. Let X and Y be a C-convex and respectively a B-convex space. Show that the quotient space X/Y is C-convex.

EXERCISE 5.3.5. Without using the results of the present section, show that every C-convex space has a cotype.

EXERCISE 5.3.6. Let X be a B-convex space. Show that $l_2(X)$ is also B-convex (here $l_2(X)$ denotes the space of sequences $x = (x_1, x_2, \dots)$, $x_k \in X$ for which $\|x\| = \left(\sum_{k=1}^\infty \|x_k\|^2\right)^{1/2} < \infty$, equipped with the norm $\|x\|$).

EXERCISE 5.3.7. Let X be a C-convex space. Show that $l_2(X)$ is also C-convex.

DEFINITION 5.3.4. The function

$$\delta_X(t) = \inf\left\{1 - \frac{\|x+y\|}{2} : x, y \in B(X),\ \|x-y\| \geq t\right\},$$

defined for $t \in [0, 2]$, is called the *modulus of convexity of the space X*. The space X is said to be *uniformly convex* if $\delta_X(t) > 0$ for all $t > 0$.

EXERCISE 5.3.8. Prove that the spaces L_p are uniformly convex for any p with $1 < p < \infty$.

EXERCISE 5.3.9. Let X be a uniformly convex space. Show that for any unconditionally convergent series $\sum_{k=1}^\infty x_k$ in X the numerical series $\sum_{k=1}^\infty \delta_X(\|x_k\|)$ converges.

EXERCISE 5.3.10. Show that any uniformly convex space is reflexive.

CHAPTER 6
SOME RESULTS FROM THE GENERAL THEORY OF BANACH SPACES

To continue our exposition, in particular, to prove the theorem of V. M. Kadets on the existence, in any infinite-dimensional Banach space, of series with nonlinear sum range, we need two theorems on the structure of infinite-dimensional spaces: Dvoretzky's theorem on almost-Euclidean sections and Mazur's theorem on basic sequences. Since these deep results are not incorporated in the standard functional analysis courses, their proof will be provided here in detail for the reader's convenience.

§1. Fréchet Differentiability of Convex Functions

The definitions and results collected in this section are well known, but, unfortunately, they are seldom included in the university analysis courses. For this reason the authors found themselves in a difficult position: not including these results would make the understanding of the already complicated next section even more difficult, while including them would mean overlapping known textbooks. We decided for the following method of exposition: give all statements without proofs, but in the order in which the reader, with a certain amount of effort, will be able to reconstruct the proofs on his own. A detailed exposition can be found, for example, in [77].

Throughout this section f will denote a real-valued function, defined on the Banach space X. Whenever we write $X = \mathbf{R}^n$ we mean that X is the n-dimensional coordinate space, equipped with some norm. In this case f will alternatively be regarded as a function of n-variables: $f(x) = f(x_1, \dots, x_n)$.

DEFINITION 6.1.1. The function f is said to be *convex* if the inequality $f(\lambda x + \mu x) \leq \lambda f(x) + \mu f(y)$ holds for all $x, y \in X$ and all $\lambda, \mu \in \mathbf{R}^+$ such that $\lambda + \mu = 1$.

It is clear that if f is convex and λ_k are positive numbers satisfying $\sum_{k=1}^{n} \lambda_k = 1$, then

$$f\left(\sum_{k=1}^{n} \lambda_k y_k\right) \leq \sum_{k=1}^{n} \lambda_k f(y_k).$$

LEMMA 6.1.1. *Let $X = \mathbf{R}^n$, and let f be a convex function that is differentiable at zero in each variable and satisfies $f(0) = 0$, $(\partial f/\partial x_k)(0) = 0$, $k = 1, \dots, n$. Then*

$$\lim_{y \to 0} \frac{f(y)}{\|y\|} = 0.$$

DEFINITION 6.1.2. The function f is said to be *Fréchet differentiable at the point* $x \in X$ if there exists a linear functional $\nabla f(x) \in X^*$ such that

$$f(z) - f(x) = \langle \nabla f(x), z - x \rangle + o(\|z - x\|),$$

or, in other words, such that

$$\lim_{z \to x} \frac{f(z) - f(x) - \langle \nabla f(x), z - x \rangle}{\|z - x\|} = 0.$$

In this case the element $\nabla f(x) \in X^*$ is called the *Fréchet differential* (*derivative*) of the function f at the point x.

It is readily verified that this definition is correct, in the sense that a function cannot have two different Fréchet derivatives as a given point.

THEOREM 6.1.1. *Let $X = \mathbf{R}^n$. Then if the convex function f is differentiable with respect to each varible at the point $x \in X$, then f is Fréchet differentiable at x and*

$$\nabla f(x) = \left(\frac{\partial f}{\partial x_1}(x), \dots, \frac{\partial f}{\partial x_n}(x) \right).$$

(Hint: reduce this assertion to Lemma 6.1.1.)

THEOREM 6.1.2. *Let $X = \mathbf{R}^n$ and let f be a convex function. Then f is Fréchet differentiable almost everywhere (with respect to the Lebesgue measure).*

THEOREM 6.1.3. *Let $\{f_n\}_{n=1}^{\infty}$ be a sequence of convex functions that converges uniformly to f on some open set D. Suppose that all f_n and f are Fréchet differentiable at the point $x_0 \in D$. Then*

$$\lim_{n \to \infty} \|\nabla f_n(x_0) - \nabla f(x_0)\| = 0.$$

REMARK 6.1.1. Suppose that some norm p on the space X is Fréchet differentiable at a point x_0. Then

$$\langle \nabla p(x_0), y \rangle \le p(y)$$

for all $y \in X$. Indeed,

$$\langle \nabla p(x_0), y \rangle = \lim_{\lambda \to 0} \frac{p(x_0 + \lambda y) - p(x_0)}{\lambda} \le \lim_{\lambda \to 0} \frac{p(x_0) + \lambda p(y) - p(x_0)}{\lambda} = p(y).$$

REMARK 6.1.2. Many of the properties of ordinary differentiability remain valid for Fréchet differentiability. In particular, the rule for differentiating composite functions holds: if $a(t)$ is a mapping of the segment $[0,1]$ into X, $a(t)$ is differentiable at a point t_0, $f(x)$ is a real-valued function on X, and $f(x)$ is Fréchet differentiable at the point $a(t_0)$, then the function $f(a(t))$ is differentiable at t_0 and

$$[f(a(t_0))]' = \langle \nabla f(a(t_0)), a'(t_0)\rangle.$$

EXERCISE 6.1.1. Show that without the convexity assumption on f Theorem 6.1.1 is false.

EXERCISE 6.1.2. Let f be a convex function on X and $x_0 \in X$. Show that there exists a functional $g \in X^*$ such that $\langle g, y - x_0\rangle \leq f(y) - f(x_0)$ for all $y \in X$. (Such a g is called a *supporting functional* of f at the point x_0.) Prove that if the convex function f is Fréchet differentiable at the point x_0, then $\nabla f(x_0)$ is the unique supporting functional of f at x_0.

EXERCISE 6.1.3. Let $1 < p < \infty$, $X = l_p$, $f(x) = \|x\|$. Show that f is Fréchet differentiable at every point $x \neq 0$. Calculate $\nabla f(x)$. Show that in the spaces l_1 and c_0 the norm is not Fréchet differentiable.

DEFINITION 6.1.3. The *modulus of smoothness* of the Banach space X is the function $\rho_X(t)$, defined by the formula

$$\rho_X(t) = \sup\left\{\frac{\|x+ty\| + \|x-ty\|}{2} - 1 \colon x, y \in S(X)\right\}.$$

The space X is said to be *uniformly smooth* if $\rho_X(t)/t \to 0$ when $t \to 0$.

EXERCISE 6.1.4. Show that in a uniformly smooth space the norm is Fréchet differentiable at every point $x \neq 0$.

EXERCISE 6.1.5. Show that the spaces $L_p(\Omega, \mu)$ with $1 < p < \infty$ are uniformly smooth.

EXERCISE 6.1.6. Let X be a uniformly smooth space and let $\sum_{k=1}^{\infty} x_k$ be a conditionally convergent series in X. Show that if $\sum_{k=1}^{\infty} \rho_X(\|x_k\|) < \infty$, then the assertion of Steinitz's theorem holds true for $\sum_{k=1}^{\infty} x_k$.

EXERCISE 6.1.7. Let the series $\sum_{k=1}^{\infty} x_k$ in X be perfectly divergent. Then $\sum_{k=1}^{\infty} \rho_X(\|x_k\|) = \infty$.

§2. Dvoretzky's Theorem

The aim of this section is to prove that a Hilbert space is finitely representable in any infinite-dimensional Banach space. To understand the proof the reader must be familiar with the theory of integration of functions of several variables at the level found in the modern analysis courses. In the sequel we shall use the result obtained, but not the method for proving it.

The main idea of the proof is to introduce a certain auxiliary measure of closeness of an infinite-dimensional space to a Euclidean space (Definition 6.2.2). This measure of closeness proves to be more convenient in the kind of problems considered here than the Banach-Mazur distance.

DEFINITION 6.2.1. A norm p given on the coordinate space X is sait to belong to the *class* P_n if $\dim X = n$ (i.e., $X = \mathbf{R}^n$) and the inequality $\frac{1}{2}\|x\|_\infty \leq p(x) \leq \|x\|_2$ holds for all $x \in X$. (Recall that $\|\cdot\|_p$ denotes the norm of the space $l_p^{(n)}$.)

LEMMA 6.2.1. *Let X be a finite-dimensional normed space, $\dim X = n \geq 4m^2$. Then there exists an m-dimensional subspace $Y \subset X$ with a basis $\{e_k\}_{k=1}^m$, such that, for any $x = \sum_{k=1}^m t_k e_k$,*

$$\frac{1}{2}\|(t_k)_1^m\|_\infty \leq \|x\|_X \leq \|(t_k)_1^m\|_2$$

(in other words, Y belongs to the class P_m).

PROOF. In the course of the proof of the Dvoretzky-Rogers lemma (Lemma 4.1.1) we established the existence of a basis $\{e_k\}_{k=1}^n$ and of elements $\{x_k\}_{k=1}^n$, such that if one regards X as a coordinate space with basis $\{e_k\}_{k=1}^n$, then $\|\cdot\|_X \leq \|\cdot\|_2$ and the following conditions are satisfied:

(A) $\|x_i\|_X = \|x_i\|_2 = 1, \quad i = 1, \ldots, n;$

(B) $x_i = \sum_{j=1}^i a_{i,j} e_j, \quad i = 1, \ldots, n;$

(C) $\sum_{k=1}^{i-1} a_{i,k}^2 = 1 - a_{i,i}^2 \leq \frac{i-1}{n}, \quad i = 1, \ldots, n.$

Recall that a *supporting functional of the ball B at a point x* is a functional x^* such that $\langle x^*, x\rangle \leq \|y\|_B$ for all $y \in X$ (here $\|\cdot\|_B$ denotes the norm in which B is the unit ball). The existence of a supporting functional is guaranteed by the Hahn-Banach theorem. Generally speaking, such a functional is not unique.

Let x_i^* be a supporting functional of the unit ball of X at the point x_i. Since $\|\cdot\|_X \leq \|\cdot\|_2$ and $\|x_i\|_X = \|x_i\|_2$, x_i^* is also a supporting functional of the unit ball of $l_2^{(n)}$. Hence, x_i^* is the functional defined by the $l_2^{(n)}$-inner product with the element x_i: $x_i^*(x) = \langle x_i, x\rangle$ for all $x \in X$. Consequently, since $|x_i^*(x)| \leq \|x\|_X$, one has that $|\langle x_i, x\rangle| \leq \|x\|_X$ for all $x \in X$.

Now take for Y the linear span of the vectors $\{e_k\}_{k=1}^m$. Let $x = \sum_{k=1}^m t_k e_k$ and $\|x_k\|_X = 1$. By construction, $\|x\|_X \leq \|(t_k)_1^m\|_2$. It remains to show that $\frac{1}{2}\|(t_k)_1^m\|_\infty \leq \|x\|_X$, i.e., that $|t_k| \leq 2$ for all $k \leq m$. We proceed by induction. Since $x_1 = a_{1,1}e_1$ and $|a_{1,1}| = 1$, we have

$$|t_1| = |\langle x, e_1\rangle| = |\langle x, x_1\rangle \leq \|x\|_X = 1.$$

Further, suppose one knows that $|t_k| \leq 2$ for any $k \leq i$. Then

$$|t_{i+1}|^2 = \frac{\left|\sum_{k=1}^{i+1} a_{i+1,k}t_k - \sum_{k=1}^i a_{i+1,k}t_k\right|^2}{|a_{i+1,i+1}|^2} \leq$$

$$\leq \frac{\left(|\langle x_{i+1}, x\rangle| + \sum_{k=1}^{i} |a_{i+1,k}t_k|\right)^2}{1 - \sum_{k=1}^{i} |a_{i+1,k}|^2} \leq \frac{\left(1 + 2\sum_{k=1}^{i} |a_{i+1,k}|\right)^2}{1 - \frac{i}{n}}$$

$$\leq \frac{n}{n-i}\left(1 + 2i^{1/2}\left(\sum_{k=1}^{i} |a_{i+1,k}|^2\right)^{1/2}\right)^2$$

$$\leq \frac{n}{n-i}\left(1 + 2i^{1/2}\left(\frac{i}{n}\right)^{1/2}\right)^2 \leq \frac{1}{4m^2 - i}(2m + 2i)^2 \leq \frac{4(2m-1)^2}{4m^2 - m + 1} \leq 4. \quad \square$$

Let X be the Euclidean space $l_2^{(n)}$ or a subspace of $l_2^{(n)}$. Denote

$$x^{\perp} = \{y \in X : \langle x, y\rangle = 0\},$$

$$\Sigma(X) = \{(x, y) : x \in S(X),\ y \in S(x^{\perp})\}.$$

Further, denote by λ_X and σ_X rotation-invariant Borel probability measures defined on $S(X)$ and $\Sigma(X)$, respectively.

LEMMA 6.2.2. *Let $X = l_2^{(n)}$ and let f be a linear functional on X. Then*

$$\int_{S(X)} |f(x)|^2 \, d\lambda_X(x) = \frac{1}{n} \|f\|_2^2.$$

PROOF. Since the measure λ_X is invariant under rotations,

$$\int_{S(X)} |f(x)|^2 \, d\lambda_X(x) = \|f\|_2^2 \int_{S(X)} |\langle e_k, x\rangle|^2 d\lambda_X(x)$$

for any $k \leq n$. Hence,

$$\int_{S(X)} |f(x)|^2 \, d\lambda_X(x) = \frac{1}{n} \|f\|_2^2 \sum_{k=1}^{n} \int_{S(X)} |\langle e_k, x\rangle|^2 d\lambda_X(x)$$

$$= \frac{1}{n} \|f\|_2^2 \int_{S(X)} \|x\|_2^2 \, d\lambda_X(x) = \frac{1}{n} \|f\|_2^2,$$

as claimed. $\square$

LEMMA 6.2.3. *Let $f(x,y)$ be a bounded measurable function on $\Sigma(X)$, and let $\dim X = n$. Then for any $k \leq n$ one can find a subspace $E \subset X$, $\dim E = k$, such that*

$$\int_{\Sigma(X)} f(x,y)\, d\sigma_E(x,y) \leq \int_{\Sigma(X)} f(x,y)\, d\sigma_X(x,y).$$

PROOF. Let Γ denote the set of all k-dimensional subspaces of X, equipped with the natural topology. Introduce a rotation-invariant Borel probability measure γ on Γ. The expression

$$\int_\Gamma d\gamma(E) \int_{\Sigma(E)} g(x,y)\, d\sigma_E(x,y)$$

defines a rotation-invariant integral on $\Sigma(X)$; moreover,

$$\int_\Gamma d\gamma(E) \int_{\Sigma(E)} 1\, d\sigma_E(x,y) = 1$$

Thanks to the uniqueness of the rotation-invariant Borel probability measure, the following identity holds:

$$\int_\Gamma d\gamma(E) \int_{\Sigma(E)} f(x,y)\, d\sigma_E(x,y) = \int_{\Sigma(X)} f(x,y)\, d\sigma_X(x,y).$$

Assume that the assertion of the lemma is false for the given k. Then

$$\int_{\Sigma(X)} f(x,y)\, d\sigma_X(x,y) = \int_\Gamma d\gamma(E) \int_{\Sigma(E)} f(x,y)\, d\sigma_E(x,y)$$

$$> \int_\Gamma d\gamma(E) \int_{\Sigma(X)} f(x,y)\, d\sigma_X(x,y) = \int_{\Sigma_(X)} f(x,y)\, d\sigma_X(x,y).$$

The contradiction we have reached proves the lemma. □

DEFINITION 6.2.2. Let $X \subset l_2^{(N)}$, $\dim X = n$, and let another norm p be given on X. The *integral measure of closeness of the norm p to the Euclidean norm* is defined as

$$v_n(p) = \int_{\Sigma(X)} |\langle \nabla p(x), y\rangle|^2 [p(x)]^{-2} d\sigma_X(x,y).$$

Notice that $v_n(tp) = v_n(p)$, and that if p is the Euclidean norm of the space $l_2^{(N)}$ and $(x,y) \in \Sigma(X)$, then

$$\nabla p(x) = x, \quad y \in x^\perp, \quad \langle p(x), y\rangle = 0;$$

hence, for the Euclidean norm $v_n(p) = 0$.

The next lemma shows that $v_n(p)$ can be indeed regarded as a measure of how close the norm p is to an Euclidean norm.

LEMMA 6.2.4. *For any $k \in \mathbf{N}$ and any $\varepsilon > 0$ there exists a $\delta(k,\varepsilon) > 0$ such that if some norm p on the space $X = l_2^{(k)}$ obeys the condition $v_k(p) \le \delta(k,\varepsilon)$, then*

$$d\big((X, p(\cdot)), (X, \|\cdot\|_2)\big) \le 1 + \varepsilon.$$

PROOF. Suppose that the assertion of the lemma is false. Take a sequence of norms q_n on $X = l_2^{(k)}$ and an $\varepsilon > 0$ such that

$$\lim_{n\to\infty} v_k(q_n) = 0, \tag{1}$$

$$\max_{x\in S(X)} q_n(x) = 1, \qquad \min_{x\in S(X)} q_n(x) \le 1 - \varepsilon \quad \text{for all } n \in \mathbf{N}.$$

Since all the q_n have the same Lipschitz constant 1, by Arzela's theorem [45] one can choose a subsequence $\{\overline{q}_n\} \subset \{q_n\}$ which converges uniformly on $S(X)$. Denote its limit by q. Then

$$\max_{x\in S(X)} q(x) = 1, \qquad \min_{x\in S(X)} q(x) \le 1 - \varepsilon.$$

Denote by A the set of all points $x \in S(X)$ at which all the norms $\overline{q}_n$ and q are Fréchet differentiable. Since $\overline{q}_n$ and q are convex and homogeneous, $\lambda_X(A) = 1$. By (1),

$$\liminf_{n\to\infty} |\langle \nabla \overline{q}_n(x), y\rangle| = 0 \tag{2}$$

almost everywhere on $\Sigma(X)$. By Remark 6.1.1, the sequence of functionals $\nabla\overline{q}_n(x)$ is bounded in norm. Next, by Theorem 6.1.3, for $x \in A$ this sequence converges to $\nabla q(x)$. Therefore, from relation (2) it follows that the inner product $\langle q(x), y\rangle$ is equal to zero for almost all pairs $(x, y) \in \Sigma(X)$. Let

$$B = \{x \in A \colon \langle q(x), y\rangle = 0 \ \text{ for all } y \in x^{\perp}\}$$

Since the function $y \mapsto \langle q(x), y\rangle$ is continuous, $\lambda_X(B) = 1$. Let $g\colon [0,1] \to S(X)$ be a smooth curve such that $g(t) \in B$ for almost all $t \in [0,1]$. Since $\langle g(t), g(t)\rangle = 1$ for all t, it follows that $g(t) \perp g'(t)$ and, by the definition of the set B, $\langle \nabla q(g(t)), g'(t)\rangle = 0$ for almost all t. Further, by Remark 6.1.2,

$$q(g(1)) - q(g(0)) = \int_0^1 [q(g(t))]'\,dt = \int_0^1 \langle \nabla q(g(t)), g'(t))\rangle\,dt = 0.$$

Hence, the function q is constant on $S(X)$ and

$$1 = \max_{x\in S(X)} q(x) = \min_{x\in S(X)} q(x) = 1 - \varepsilon;$$

we have reached a contradiction. $\square$

LEMMA 6.2.5. *Let* P_n *be the class of norms introduced in Definition* 6.2.1. *Then* $\lim_{n\to\infty}\sup_{p\in\mathrm{P}_n} v_n(p) = 0$.

PROOF. Let $X = l_2^{(n)}$, $p \in \mathrm{P}_n$. By Lemma 6.2.2,

$$v_n(p) = \int_{S(X)} \left(\int_{S(x^\perp)} |\langle \nabla p(x), y\rangle|^2 \, d\lambda_{x^\perp}(y) \right) [p(x)]^{-2} \, d\lambda_X(x)$$

$$\leq \frac{1}{n-1} \int_{S(X)} (\|\nabla p(x)\|_2)^2 \, [p(x)]^{-2} \, d\lambda_X(x).$$

By Remark 6.1.1, $\langle \nabla p(x), y\rangle \leq p(y)$ for all y; also, $p(y) \leq \|y\|_2$ for any $p \in \mathrm{P}_n$. Consequently, $\|\nabla p(x)\|_2 \leq 1$. It follows that

$$v_n(p) \leq \frac{1}{n-1} \int_{S(X)} [p(x)]^{-2} \, d\lambda_X(x) \leq \frac{1}{n-1} \int_{S(X)} 4\left[\|x\|_\infty\right]^{-2} d\lambda_X(x) \stackrel{\text{def}}{=} a_n.$$

The last quantity does not depend on p, and so $\sup_{p\in\mathrm{P}_n} v_n(p) \leq a_n$. Let us estimate the behavior of a_n when $n \to \infty$. To do this it is convenient to ensure that the integration will be carried over the same measure space for all values of n. To this end, choose in some probability space (Ω, Σ, μ) a sequence of independent, normally distributed random variables $\{\nu_k\}_{k=1}^\infty$ with the same distribution and mean value zero. Fix n for the moment and consider the random variables $t_k = \nu_k / \left(\sum_{i=1}^n |\nu_i|^2\right)^{1/2}$. The mapping $\omega \mapsto (t_1(\omega), \ldots, t_n(\omega))$ transforms the measure μ into the measure λ_X on $S(X)$. Consequently,

$$\frac{n-1}{4n} a_n = \int_{S(X)} \frac{1}{n} \left[\|x\|_\infty\right]^{-2} d\lambda_X(x) = \frac{1}{n} \mathbf{E} \left(\max_{1\leq k\leq n} |t_k| \right)^{-2}$$

$$= \frac{1}{n} \mathbf{E} \left(\sum_{k=1}^n |\nu_k|^2 \cdot \left(\max_{1\leq k\leq n} |\nu_k| \right)^{-2} \right) = \mathbf{E} \left(|\nu_1|^2 \cdot \left(\max_{1\leq k\leq n} |\nu_k| \right)^{-2} \right).$$

Thus, we managed to reduce the problem to the estimation of mean values on a single probability space. The quantity $R_n = |\nu_1|^2 \cdot \left(\max_{1\leq k\leq n} |\nu_k|\right)^{-2}$ is monotonically nonincreasing in n. Moreover,

$$\mu \left(\left(\max_{1\leq k\leq n} |\nu_k| \right)^{-2} > A \right) = \prod_{k=1}^n \mu \left(|\nu_k| < 1/\sqrt{A} \right) = \left[\mu \left(|\nu| > 1/\sqrt{A} \right) \right]^n$$

for any $A > 0$. We see that the sequence R_n converges to 0 with probability 1. Hence, by the dominated convergence theorem, $\lim_{n\to\infty} a_n = \lim_{n\to\infty} \mathbf{E}(R_n) = 0$. □

THEOREM 6.2.1 (DVORETZKY'S THEOREM). *Let k be an arbitrary natural number and let $\varepsilon > 0$. Then there exists a number $n(k, \varepsilon)$ such that, for any normed space X with $\dim X > n(k, \varepsilon)$ there is a k-dimensional subspace $Y \subset X$ such that $d\left(Y, l_2^{(k)}\right) < 1 + \varepsilon$.*

PROOF. Using Lemma 6.2.5, put $n(k, \varepsilon) = 4m^2$, in which m is such that $\sup_{p \in \mathrm{P}_m} v_m(p) < \delta(k, \varepsilon)$, where $\delta(k, \varepsilon)$ is as in Lemma 6.2.4. By Lemma 6.2.1, there exist an m-dimensional subspace $X_1 \subset X$ and elements $\{e_i\}_{i=1}^m$ in X_1, such that

$$\frac{1}{2}\|(t_i)_1^m\|_\infty \leq \left\|\sum_{i=1}^m t_i e_i\right\|_{X_1} \leq \|(t_i)_1^m\|_2.$$

By construction, $\|\cdot\|_{X_1} \in \mathrm{P}_m$ and $v_m\left(\|\cdot\|_{X_1}\right) < \delta(k, \varepsilon)$. Then Lemma 6.2.3 yields a subspace $Y \subset X_1$ such that $v_k\left(\|\cdot\|_Y\right) \leq \delta(k, \varepsilon)$. By Lemma 6.2.4, $d\left(Y, l_2^{(k)}\right) \leq 1+\varepsilon$. □

COROLLARY. *The space l_2 is finitely representable in any infinite-dimensional Banach space.*

Let us point out that, by Exercise 5.1.3, only Hilbert space can possess this universality property.

Our exposition above followed the work of T. Figiel [21]. A detailed exposition of the beautiful modern theory connected with Dvoretzky's theorem can be found in the very useful monograph of V. D. Milman and G. Schechtman [99].

EXERCISE 6.2.1. Show that the space c_0 contains no infinite-dimensional subspaces isomorphic to l_2.

EXERCISE 6.2.2. Estimate from above the numbers $n(k, \varepsilon)$.

§3. Basic Sequences

DEFINITION 6.3.1. A sequence $\{e_k\}_{k=1}^\infty$ in an infinite-dimensional Banach space is called a *basis* if every element $x \in X$ has a unique representation as a series in the elements e_k:

$$x = \sum_{k=1}^\infty a_k(x) e_k, \quad a_k(x) \in \mathbf{R}.$$

This definition is the natural generalization of the notion of a "basis in a finite-dimensional space."

Obviously, bases defined in this manner can exist only in a separable space. All classical separable Banach spaces posess bases. The first example of a separable space that has no basis was constructed by P. Enflo in 1973 [20].

DEFINITION 6.3.2. A sequence of elements $\{e_k\}_{k=1}^{\infty}$ of a Banach space is called a *basic sequence* if $\{e_k\}_{k=1}^{\infty}$ is a basis in $\overline{\mathrm{Lin}}\,\{e_k\}_{k=1}^{\infty}$.

The next theorem is given here without a proof, for which the reader is referred to the books [55, Chapter 3, § 6] and [53, page 2].

THEOREM 6.3.1 (S. BANACH). *In order for a sequence $\{e_k\}_{k=1}^{\infty}$, $e_k \neq 0$, to be a basic sequence it is necessary and sufficient that there exist a constant $K > 0$ such that*

$$\left\| \sum_{j=1}^{m+n} a_j e_j \right\| \geq K \left\| \sum_{j=1}^{m} a_j e_j \right\| \tag{1}$$

for all $m, n \in \mathbf{N}$ and $a_j \in \mathbf{R}$. □

DEFINITION 6.3.3. The *basic constant* of the basic sequence $\{e_k\}_{k=1}^{\infty}$ is the supremum of all constants K for which inequality (1) holds.

Notice that if K is the basic constant of the sequence $\{e_k\}_{k=1}^{\infty}$, then for any $a_1, \ldots, a_n$ and any $n \in \mathbf{N}$ we have

$$\left\| \sum_{i=1}^{n} a_i e_i \right\| \geq \frac{K}{2} \max_{j \leq n} \{|a_j| \cdot \|e_j\|\}. \tag{2}$$

Indeed, for $j \leq n$ relation (1) gives

$$\left\| \sum_{i=1}^{n} a_i e_i \right\| \geq \frac{K}{2} \left(\left\| \sum_{i=1}^{j} a_i e_i \right\| + \left\| \sum_{i=1}^{j-1} a_i e_i \right\| \right) \geq \frac{K}{2} |a_j| \cdot \|e_j\|.$$

THEOREM 6.3.2 (THE KREIN-MIL'MAN-RUTMAN THEOREM). *Let $\{e_i\}_{i=1}^{\infty} \subset S(X)$ be a basic sequence with basic constant K and let $\sum_{i=1}^{\infty} \|e_i - y_i\| = K_1 < K/2$. Then $\{y_i\}_{i=1}^{\infty}$ is also a basic sequence.*

PROOF. Define an operator $T\colon \mathrm{Lin}\,\{e_i\} \to \mathrm{Lin}\,\{y_i\}$ by the rule $Te_i = y_i$, $i = 1, 2, \ldots$. To prove that $\{y_i\}_{i=1}^{\infty}$ is a basic sequence it suffices to show that T is an isomorphism of normed spaces. First let us check that T is bounded. Let $\sum_{i=1}^{n} a_i e_i$ be an arbitrary element in $\mathrm{Lin}\,\{e_i\}$ Then

$$\left\| T\left(\sum_{i=1}^{n} a_i e_i \right) \right\| = \left\| \sum_{i=1}^{n} a_i y_i \right\| = \left\| \sum_{i=1}^{n} a_i e_i + \sum_{i=1}^{n} a_i (y_i - e_i) \right\|$$

$$\leq \left\| \sum_{i=1}^{n} a_i e_i \right\| + \sum_{i=1}^{n} |a_i| \cdot \|y_i - e_i\| \leq \left\| \sum_{i=1}^{n} a_i e_i \right\| + K_1 \max_{1 \leq i \leq n} |a_i|$$

$$\leq \left\| \sum_{i=1}^{n} a_i e_i \right\| \left(1 + 2\frac{K_1}{K} \right) \leq 2 \left\| \sum_{i=1}^{n} a_i e_i \right\|.$$

Thus, we have shown that $\|T\| \leq 2$. Let us establish the boundedness of the operator T^{-1}, and hence the invertibility of T. We have

$$\left\|T\left(\sum_{i=1}^{n} a_i e_i\right)\right\| = \left\|\sum_{i=1}^{n} a_i e_i + \sum_{i=1}^{n} a_i(y_i - e_i)\right\| \geq$$

$$\geq \left\|\sum_{i=1}^{n} a_i e_i\right\| - K_1 \max_{1\leq i\leq n} |a_i| \geq \left(1 - 2\frac{K_1}{K}\right) \|\sum_{i=1}^{n} a_i e_i\|.$$

Thus, T is an isomorphism. □

The result just established asserts that the property of being a basic sequence is preserved under small perturbations.

Below we shall prove Mazur's theorem on the existence of basic sequences in any infinite-dimensional Banach space. To facilitate the exposition, let us introduce the following definition: a subspace Z is said to be *ε-orthogonal* to a subspace Y if $\|y + z\| \geq (1 - \varepsilon)\|y\|$ for all $y \in Y$ and all $z \in Z$.

LEMMA 6.3.1. *Let X be a Banach space, Y a finite-dimensional subspace of X, and Z an infinite-dimensional subspace of X. Then for any $\varepsilon > 0$ there exists a subspace $F \subset Z$ of finite codimension in Z, such that F is ε-orthogonal to Y.*

PROOF. Pick a finite ε-net $\{a_i\}_{i=1}^n$ on the unit sphere of the subspace Y and denote by f_i a supporting functional at the point a_i (i.e., $\|f_i\| = 1$ and $f_i(a_i) = 1$). Then the subspace $E = \bigcap_{i=1}^n \operatorname{Ker} f_i$ is ε-orthogonal to Y. Indeed, let $y \in Y$ and $e \in E$. Choose an element a_i such that $\|(y/\|y\|) - a_i\| < \varepsilon$. Then

$$\|y + e\| \geq f_i(y + e) = f_i(y) = \|y\| \left(f\left(\frac{y}{\|y\|} - a_i\right) + f(a_i)\right)$$

$$= \|y\| \left(1 + f_i\left(\frac{y}{\|y\|} - a_i\right)\right) \geq \|y\|(1 - \varepsilon).$$

Since E has finite codimension in X, the subspace $F = E \cap Z$ has the required properties. □

THEOREM 6.3.3. *Let X be an infinite-dimensional Banach space. Then for any $\varepsilon > 0$ there exists a basic sequence $\{e_n\}_{n=1}^\infty \subset X$ whose basic constant is not smaller than $1 - \varepsilon$.*

PROOF. Pick an arbitrary element $e_1 \in S(X)$ and put $Y_1 = \operatorname{Lin}\{e_1\}$. Lemma 6.3.1 yields a finite-codimensional subspace $F_1 \subset X$ that is ε-orthogonal to Y_1. Then pick an arbitrary element $e_2 \in S(F_1)$, put $Y_2 = \operatorname{Lin}\{e_1, e_2\}$, and take a subspace $F_2 \subset F_1$ of finite codimension in F_1 that is ε-orthogonal to Y_2. Continuing this process, we obtain a sequence of elements $\{e_n\}_{n=1}^\infty$ and sequences of subspaces $Y_n = \operatorname{Lin}\{e_k\}_{k=1}^n$ and $F_n \supset \{e_k\}_{k=n+1}^\infty$, such that F_n is ε-orthogonal to Y_n. Let us verify that $\{e_n\}_{n=1}^{\infty}$ is a basic sequence. Let $\{a_i\}_{i=1}^{n+m}$ be arbitrary numbers. Then $y = \sum_{i=1}^{n} a_i e_i \in Y_n$, $z = \sum_{i=n+1}^{n+m} a_i e_i \in F_n$, and

$$\left\| \sum_{i=1}^{n+m} a_i e_i \right\| = \|y + z\| \geq (1 - \varepsilon)\|y\| = (1 - \varepsilon) \left\| \sum_{i=1}^{n} a_i e_i \right\|. \qquad \square$$

DEFINITION 6.3.4. A Banach space X is said to be *finitely saturated* by a Banach space Y (and one writes $Y \overset{\mathrm{f}}{\Rightarrow} X$) if Y is finitely representable in any finite-codimensional subspace of X.

DEFINITION 6.3.5. A sequence $\{X_i\}_{i=1}^{\infty}$ of subspaces of the space X is called a *basic sequence of subspaces* if there exists a constant $K > 0$ such that the inequality

$$\left\| \sum_{i=1}^{n+m} x_i \right\| \geq K \left\| \sum_{i=1}^{n} x_i \right\|$$

holds for all $n, m \in \mathbf{N}$ and all $x_i \in X_i$, $i = 1, 2, \ldots, n + m$. The supremum of all such constants is called the *basic constant* of the sequence $\{x_i\}_{i=1}^{\infty}$.

The next theorem will play an essential role in further applications.

THEOREM 6.3.4. *Let X and Y be Banach spaces, $Y \overset{\mathrm{f}}{\Rightarrow} X$ and let $\{\overline{Y}_i\}_{i=1}^{\infty}$ be an arbitrary sequence of finite-dimensional subspaces of Y. Then for any $\varepsilon > 0$ there exists a basic sequence of subspaces $\{X_i\}_{i=1}^{\infty}$ in X with basic constant $1 - \varepsilon$, such that $d(X_i, \overline{Y}_i) \leq 1 + \varepsilon$ for all i.*

PROOF. We argue in much the same way as in the proof of Theorem 6.3.3, with the difference that instead of an element $e_n \in F_{n-1}$ we must choose a subspace $X_n \subset F_{n-1}$ such that $d(X_n, \overline{Y}_n) < 1 + \varepsilon$, and then define Y_n as $\operatorname{Lin} \bigcup_{i=1}^{n} X_i$. $\square$

REMARK 6.3.1. If $\{X_i\}_{i=1}^{\infty}$ is a basic sequence of subspaces with basic constant K and if $x_i \in X_i$, $i \in \mathbf{N}$, then by analogy with inequality (2) one can readily show that

$$\left\| \sum_{i=1}^{n} x_i \right\| \geq \frac{K}{2} \max_{1 \leq i \leq n} \|x_i\|.$$

EXERCISE 6.3.1. Show that the spaces c_0, $C[0,1]$, and l_p, $L_p[0,1]$, with $1 \leq p < \infty$, have bases.

§4. Some Applications to Conditionally Convergent Series

In this section we collected a number of interesting facts that are somewhat beyond the general scope of our exposition. The proofs of these results rely on the technique of singling out basic sequences.

THEOREM 6.4.1. *In any infinite-dimensional Banach space there are series whose sum range reduces to a single point, but which are not unconditionally convergent.*

PROOF. Let $\{e_n\}_{n=1}^{\infty}$ be a basic sequence in the space X, $\|e_n\| = 1$, $n = 1, 2 \dots$. The series

$$e_1 - e_1 + \frac{1}{2}e_2 - \frac{1}{2}e_2 + \frac{1}{2}e_2 - \frac{1}{2}e_2 + \frac{1}{4}e_3 - \frac{1}{4}e_3 + \dots, \tag{1}$$

where each group of 2^n terms of the form $\pm 2^{-n+1}e_n$ mutually annihilate, converges to zero. If the terms of the series (1) are rearranged so that it remains convergent, then its sum will be again zero, since the projection on any vector in the basis is zero. At the same time, if in (1) we replace all signs $\pm$ by $+$, we obtain a divergent series, because the series segments formed by the vectors $2^{-n-1}e_n$ give the vector $2e_n$, which does not converge to zero, contrary to the Cauchy criterion. □

Notice that series (1) obeys the assertion of Steinitz's theorem: $\mathrm{SR}(\sum_{k=1}^{\infty} x_k) = 0 + \Gamma^{\perp}$; indeed, here the set of absolute convergence functionals includes the coordinate functionals as well as the functionals that annihilate all vectors e_n. Below we shall consider series for which any bounded functional is an absolutely convergence functional.

DEFINITION 6.4.1. A series $\sum_{k=1}^{\infty} x_k$ of elements of a Banach space X is said to be *weakly absolutely convergent* if $\sum_{k=1}^{\infty} |f(x_k)| < \infty$ for all $f \in X^*$.

Let us remark that an unconditionally convergent series is taken by any continuous linear functional into an unconditionally convergent numerical series; and since for numerical series unconditional and absolute convergence coincide, every unconditionally convergent series is weakly absolutely convergent.

EXAMPLE 6.4.1. In the space c_0 consider the series $\sum_{k=1}^{\infty} e_k$ formed by the unit vectors of the standard basis. This series diverges, since its general term does not converge to zero. Take an arbitrary linear functional $f \in l_1 = (c_0)^*$; then $f = (f_1, f_2, \dots)$, $f_k \in \mathbf{R}$, $\sum_{k=1}^{\infty} |f_k| = \|f\| < \infty$. We have $\sum_{k=1}^{\infty} f(e_k) = \sum_{k=1}^{\infty} |f_k| < \infty$, i.e., our series is weakly absolutely convergent. Thus, we constructed a series in c_0 that is weakly absolutely convergent, but diverges in the norm topology. Morevoer, this series diverges in the weak topology as well: if it were convergent, the coordinates of its sum would all be equal to 1, but c_0 contains no such element.

COROLLARY 6.4.1. *If the Banach space X contains an isomorphic copy of c_0, then in X there exists a weakly absolutely convergent series which diverges (in the strong as well as in the weak topology).*

As we shall see below, in spaces that contain no isomorphic copy of c_0, weak absolute convergence of a series implies its unconditional convergence. To prove this theorem of Bessaga and Pełczyński we need the following assertions, which are of independent interest.

LEMMA 6.4.1. *Let the series $\sum_{k=1}^{\infty} x_k$ be weakly absolutely convergent. Then there exists a constant $A > 0$ such that $\sum_{k=1}^{\infty} |f(x_k)| \leq A\|f\|$ for all $f \in X^*$.*

PROOF. Let $\{e_k\}_{k=1}^{\infty}$ be the standard basis of the space l_1. Define operators $S_n\colon X^* \to l_1$ by the formula $S_n(f) = \sum_{k=1}^{n} f(x_k)e_k$. These operators are continuous and, for any fixed $f \in X^*$, the numbers $\|S_n(f)\| = \sum_{k=1}^{n} |f(x_k)|$ are bounded by $\sum_{k=1}^{\infty} |f(x_k)|$. By the Banach-Steinhaus theorem, this means that $\sup_n \|S_n\| = A < \infty$. Therefore, $\sum_{k=1}^{\infty} |f(x_k)| = \sup_n \|S_n(f)\| \leq A\|f\|$. □

LEMMA 6.4.2. *Let X be a Banach space, Y a finite-dimensional subspace of X, and $\{x_i\}_{i=1}^{\infty}$ a sequence of elements of X such that $x_i \to 0$ weakly and $\inf_i \|x_i\| = C > 0$. Then for any $\varepsilon > 0$ there exists a number n such that the subspace $\operatorname{Lin}\{x_n\}$ is ε-orthogonal to Y.*

PROOF. Pick a finite $\varepsilon/2$-net $\{y_i\}_{i=1}^{N}$ on $S(Y)$ and denote by f_i a supporting functional at the point y_i. Take $\delta < C\varepsilon/4$ and choose a number n such that $\max_{1\leq i\leq N} |f_i(x_n)| < \delta$. We claim that n has the desired property. Indeed, let $y \in Y$ and $a \in \mathbf{R}$. We need to show that $\|y + ax_n\| \geq (1-\varepsilon)\|y\|$. Thanks to homogeneity, it suffices to prove the last inequality for $y \in S(Y)$. If $|a| \geq 2/\|x_n\|$, then

$$\|y + ax_n\| \geq \|ax_n\| - \|y\| \geq 2 - 1 = 1.$$

Hence, we may assume that $|a| < 2/\|x_n\|$. Pick an element y_k such that $\|y - y_k\| < \varepsilon/2$. Then

$$\|y + ax_n\| \geq \|y_k + ax_n\| - \frac{\varepsilon}{2} \geq f_k(y_k + ax_n) - \frac{\varepsilon}{2}$$

$$= 1 - \frac{\varepsilon}{2} + af_k(x_n) \geq 1 - \frac{\varepsilon}{2} - |a|\delta \geq 1 - \frac{\varepsilon}{2} - \frac{2\delta}{\|x_n\|} \geq 1 - \frac{\varepsilon}{2} - \frac{2\delta}{C} \geq 1 - \varepsilon. \quad \square$$

THEOREM 6.4.2. *Let $\{x_i\}_{i=1}^{\infty}$ be a sequence of elements of a Banach space X such that $x_i \to 0$ weakly and $\inf_i \|x_i\| > 0$. Then for any $\varepsilon > 0$ there exists a subsequence $\{y_i\}$ of $\{x_i\}$ which is a basic sequence with basic constant larger than or equal to $1-\varepsilon$.*

PROOF. Let $\varepsilon_i > 0$ be numbers such that $\prod_{i=1}^{\infty}(1-\varepsilon_i) > 1-\varepsilon$. Set $y_1 = x_1$. Next, using the preceding lemma, pick successively $y_2, y_3, \dots \in \{x_1, x_2, \dots\}$ such that $\operatorname{Lin}\{y_{n+1}\}$ will be ε_n-orthogonal to $\operatorname{Lin}\{y_1, \dots, y_n\}$ for each n. Then, for any coefficients a_i and any $n, m \in \mathbf{N}$,

$$\left\|\sum_{i=1}^{n+m} a_i y_i\right\| \geq (1-\varepsilon_{n+m-1})\left\|\sum_{i=1}^{n+m-1} a_i y_i\right\|$$

$$\geq (1-\varepsilon_{n+m-2})(1-\varepsilon_{n+m-1})\left\|\sum_{i=1}^{n+m-2} a_i y_i\right\| \geq \dots \geq \prod_{i=1}^{\infty}(1-\varepsilon_i)\left\|\sum_{i=1}^{n} a_i y_i\right\|. \quad \square$$

THEOREM 6.4.3 (THE BESSAGA-PEŁCZYŃSKI THEOREM). *The following assertions are equivalent:*

(a) *The Banach space X contains no subspaces isomorphic to c_0.*

(b) *Every weakly absolutely convergent series in X is weakly convergent.*

(c) *Every weakly absolutely convergent series in X is unconditionally convergent.*

(d) *Every weakly absolutely convergent series in X is norm convergent.*

PROOF. (d) $\Rightarrow$ (c). Let the series $\sum_{i=1}^{\infty} x_i$ in X be weakly absolutely convergent. Then all series $\sum_{i=1}^{\infty} \alpha_i x_i$, $\alpha_i = \pm 1$, are weakly absolutely convergent as well. By (d), this means that all series $\sum_{i=1}^{\infty} \alpha_i x_i$ are norm-convergent, which means precisely that the series $\sum_{i=1}^{\infty} x_i$ is unconditionally convergent.

The implication (c) $\Rightarrow$ (b) is obvious, while the implication (b) $\Rightarrow$ (a) follows from Corollary 6.4.1. It remains to prove that (a) $\Rightarrow$ (d). Suppose that in X there exists a divergent series which is weakly absolutely convergent. Using the fact that the series is divergent one can group its terms so that the resulting series with terms $y_k = \sum_{i=n_k+1}^{n_{k+1}} x_i$, $0 = n_1 < n_2 < \dots$ will satisfy the condition $\inf_k \|y_k\| = \delta > 0$. The series $\sum_{k=1}^{\infty} y_k$ is again weakly absolutely convergent. Consequently, the sequence $\{y_k\}$ converges weakly to zero. Extract from $\{y_k\}$ a basic subsequence $\{z_k\}$. The series $\sum_{k=1}^{\infty} z_k$ is also weakly absolutely convergent and, by Lemma 6.4.1, there exists a constant $A > 0$ such that $\sum_{k=1}^{\infty} |f(z_k)| \leq A\|f\|$ for all $f \in X^*$. Define an operator $T\colon c_0 \to X$ by the rule $Te_k = z_k$, where $\{e_k\}$ is the standard basis of c_0. Let $t_1, \dots, t_n$ be arbitrary numbers. Then

$$\left\| T\left(\sum_{k=1}^{n} t_k e_k\right)\right\| = \left\|\sum_{k=1}^{n} t_k z_k\right\| = \sup_{f\in S(X^*)} \sum_{k=1}^{n} t_k f(z_k)$$

$$\leq \max_k |t_k| \cdot \sup_{f\in S(X^*)} \sum_{k=1}^{n} |f(z_k)| \leq A \max_k |t_k| = A \left\|\sum_{k=1}^{n} t_k e_k\right\|.$$

Hence, the operator T is well defined and continuous. Let us show that T is an isomorphic embedding of c_0 into X. By Remark 6.3.1, there exists a number $K > 0$ such that

$$\left\|\sum t_k z_k\right\| \geq K \max_k \|t_k z_k\| \geq K\delta \max_k |t_k|.$$

Consequently,

$$\left\| T\left(\sum_{k=1}^{n} t_k e_k\right)\right\| = \left\|\sum_{k=1}^{n} t_k z_k\right\| \geq C\delta \left\|\sum_{k=1}^{n} t_k e_k\right\|. \qquad \square$$

EXERCISE 6.4.1. Show that a series $\sum_{k=1}^{\infty} x_k$ is weakly absolutely convergent if and only if

$$\sup\left\{\left\|\sum_{k=1}^{n} \alpha_k x_k\right\| : n \in \mathbf{N},\ \alpha_k = \pm 1\right\} < \infty.$$

EXERCISE 6.4.2. Suppose $\lim_{k\to\infty}\|x_k\| = 0$ and the series $\sum_{k=1}^{\infty} x_k$ is weakly absolutely convergent. Does this imply that the series is convergent?

DEFINITION 6.4.2. A series $\sum_{k=1}^{\infty} x_k$ is said to be *weakly unconditionally convergent* if the series $\sum_{k=1}^{\infty} \alpha_k x_k$ is weakly convergent for any choice of signs $\alpha_k = \pm 1$.

EXERCISE 6.4.3. Proceeding by analogy with the proof of Theorem 6.4.3, prove Orlicz's theorem on the coincidence of the notions of weak unconditional convergence and ordinary unconditional convergence.

CHAPTER 7
STEINITZ'S THEOREM AND B-CONVEXITY

The main results presented in this chapter were obtained in the period 1984–1986.

§1. Conditionally Convergent Series in Spaces with Infratype

In Chapter 2 we presented Pecherskii's theorem, and all further results on the linearity of the sum range were derived in this book from that theorem. Among such results is Chobanyan's theorem (Theorem 3.4.1): In a space of type p the condition $\sum_{k=1}^{\infty} \|x_k\|^p < \infty$ is sufficient in order for the assertion of Steinitz's theorem to hold for the sum range $\mathrm{SR}(\sum_{k=1}^{\infty} x_k)$. In this section we shall strengthen Theorem 3.4.1, replacing the assumption that the space has type p by the assumption that it has infratype p (see Definition 5.3.2). This result of V. M. Kadets and M. I. Ostrovskii, obtained about the same time as Theorem 3.4.1, is one of the few that up to now was not shown to be deducible from Pecherskii's theorem.

In accordance with Lemma 2.3.1, we only need to establish some analogue of the Permutation Lemma (in order to obtain condition (A)) and some analogue of the Rounding-off-Coefficients Lemma (to obtain condition (B)).

DEFINITION 7.1.1. A space X is said to have *property* A_1 *with exponent* p *and constant* C if for any $n \in \mathbf{N}$ and any collection $\{x_i\}_{i=1}^n$ of elements of X there exists a permutation σ of the set $\{1, \dots, n\}$ such that

$$\max_{j \le n} \left\| \sum_{k=1}^{j} x_{\sigma(k)} - \frac{j}{n} \sum_{k=1}^{n} x_k \right\| \le C \left(\sum_{k=1}^{n} \|x_k\|^p \right)^{1/p}.$$

Condition A_1 is an analogue of the Steinitz lemma. Clearly, in a space with property A_1 from the assumption that $\sum_{k=1}^{\infty} \|x_k\|^p < \infty$ it follows that the series $\sum_{k=1}^{\infty} x_k$ satisfies condition (A) of Lemma 2.3.1. The aim of the following arguments is to show that property A_1 is equivalent to the existence of an infratype.

DEFINITION 7.1.2. A space X is said to have *property* A_2 *with exponent* p *and constant* C if for any $m \in \mathbf{N}$ and any collection $\{x_i\}_{i=1}^{2m}$ of elements of X there exists a permutation σ such that

$$\left\| \sum_{k=1}^{m} x_{\sigma(k)} - \frac{1}{2} \sum_{k=1}^{2m} x_k \right\| \le C \left(\sum_{k=1}^{m} \|x_k\|^p \right)^{1/p}.$$

Obviously, property A_1 implies property A_2 (it suffices to replace n by $2m$ and j by m in Definition 7.1.1). Furthermore, since

$$\left\| \sum_{k=1}^{n} x_{\sigma(k)} - \frac{1}{2} \sum_{k=1}^{2n} x_k \right\| = \frac{1}{2} \left\| \sum_{k=1}^{n} x_{\sigma(k)} - \sum_{k=n+1}^{2n} x_{\sigma(k)} \right\|,$$

property A_2 with exponent p implies that the space has infratype p.

LEMMA 7.1.1. *If the space X has infratype p with constant C, then it has property A_1 with the same parameters.*

PROOF. Let $\{x_i\}_{i=1}^{2m}$ be an arbitrary collection of elements of X. Consider the auxiliary elements $y_k = x_{2k} - x_{2k-1}$ and choose $\alpha_k = \pm 1$ so that

$$\left\| \sum_{k=1}^{m} \alpha_k y_k \right\| \leq C \left(\sum_{k=1}^{m} \|y_k\|^p \right)^{1/p}.$$

Then

$$\left\| \sum_{k=1}^{m} \alpha_k y_k \right\| \leq C \left(\sum_{k=1}^{m} (\|x_{2k}\| + \|x_{2k-1}\|)^p \right)^{1/p}$$

$$\leq C \left(\sum_{k=1}^{m} \|x_{2k}\|^p \right)^{1/p} + C \left(\sum_{k=1}^{m} \|x_{2k-1}\|^p \right)^{1/p}$$

$$\leq C \left(\sum_{k=1}^{2m} \|x_k\|^p \right)^{1/p} + C \left(\sum_{k=1}^{2m} \|x_k\|^p \right)^{1/p} = 2C \left(\sum_{k=1}^{2m} \|x_k\|^p \right)^{1/p}.$$

On the other hand,

$$\sum_{k=1}^{m} \alpha_k y_k = \sum_{k=1}^{m} \alpha_k x_{2k} - \sum_{k=1}^{m} \alpha_k x_{2k-1} = \sum_{i=1}^{2m} \beta_i x_i,$$

where $\beta_i = \pm 1$ and there are as many values of i for which $\beta_i = 1$ as there are for which $\beta_i = -1$. Hence, if the permutation σ is taken so that $\beta_{\sigma(k)} = 1$ for $k = 1, \ldots, n$, then

$$\left\| \sum_{k=1}^{m} x_{\sigma(k)} - \frac{1}{2} \sum_{k=1}^{2m} x_k \right\| = \frac{1}{2} \left\| \sum_{k=1}^{2m} \beta_k x_k \right\| \leq C \left(\sum_{k=1}^{2m} \|x_k\|^p \right)^{1/p}. \qquad \square$$

LEMMA 7.1.2. *Suppose that X has property A_2 with exponent p and constant C. Then for any $n \in \mathbf{N}$ and any collection of elements $\{x_k\}_{k=1}^{2^n} \subset X$ there exists a permutation σ such that, for each $j \in \{1, 2, \ldots, 2^n\}$,*

$$\left\| \sum_{k=1}^{j} x_{\sigma(k)} - \frac{j}{2^n} \sum_{k=1}^{2^n} x_k \right\| \leq C \left(\sum_{k=1}^{2^n} \|x_k\|^p \right)^{1/p} \cdot \left(\frac{4}{1-\varepsilon} - \frac{j}{2^{n-1}} - 2 \right), \qquad (1)$$

where $\varepsilon = (1/2)^{1/p}$.

PROOF. We proceed by induction on n. For $n = 1$ the assertion of the lemma is true. Suppose it is true for $n = N - 1$, and let us show that it remains true for $n = N$. Pick an arbitrary collection $\{x_k\}_{k=1}^{2^n}$ of elements of X. Using the inductive hypothesis and property A_2 one can construct a permutation σ with the following properties:

(a) $\left\| \sum_{k=1}^{2^{N-1}} x_{\sigma(k)} - \frac{1}{2} \sum_{k=1}^{2^N} x_k \right\| \leq C \left(\sum_{k=1}^{2^N} \|x_k\|^p \right)^{1/p}$;

(b) $\left\| \sum_{k=2^{N-1}+1}^{2^N} x_{\sigma(k)} \right\|^p \leq \frac{1}{2} \sum_{k=1}^{2^N} \|x_k\|^p$.

(c) for any $j \in \{1, \ldots, 2^{N-1}\}$,

$$\left\| \sum_{k=1}^{j} x_{\sigma(k)} - \frac{1}{2^{N-1}} \sum_{k=1}^{2^{N-1}} x_{\sigma(k)} \right\| \leq C \left(\sum_{k=1}^{2^{N-1}} \|x_{\sigma(k)}\|^p \right)^{1/p} \cdot \left(\frac{4}{1-\varepsilon} - \frac{j}{2^{N-2}} - 2 \right);$$

(d) for any j, $2^{N-1} < j < 2^N$,

$$\left\| \sum_{k=2^{N-1}+1}^{j} x_{\sigma(k)} - \frac{j - 2^{N-1}}{2^{N-1}} \sum_{k=2^{N-1}+1}^{2^N} x_{\sigma(k)} \right\|$$

$$\leq C \left(\sum_{k=2^{N-1}+1}^{2^N} \|x_{\sigma(k)}\|^p \right)^{1/p} \cdot \left(\frac{4}{1-\varepsilon} - \frac{j}{2^{N-2}} - 2 \right).$$

Let us explain the order in which the indicated permutation σ is constructed. Property A_2 yields a partition of the index set $\{1, \ldots, 2^N\}$ into subsets I_1 and I_2, each consisting of 2^{N-1} elements, such that

$$\frac{1}{2} \left\| \sum_{i \in I_1} x_i - \sum_{i \in I_2} x_i \right\| \leq C \left(\sum_{k=1}^{2^N} \|x_k\|^p \right)^{1/p}.$$

From the sets I_1 and I_2 pick the one (say, I_2) for which $\sum_{i \in I} \|x_i\|^p \leq \frac{1}{2} \sum_{i=1}^{2^N} \|x_i\|^p$. To obtain the desired permutation σ, write first the elements $\{x_i\}_{i \in I_1}$, permuted in accordance with the inductive hypothesis, and then the elements $\{x_i\}_{i\in I_2}$, also permuted in accordance with the inductive hypothesis. For this construction properties (a)–(d) are automatically satisfied.

Let us verify that inequality (1) holds for the permutation σ constructed above. Consider first the case $j < 2^{N-1}$. Using the triangle inequality and then properties (a) and (c), we obtain the following estimates:

$$\left\|\sum_{k=1}^{j} x_{\sigma(k)} - \frac{j}{2^N}\sum_{k=1}^{2^N} x_k\right\| \le \left\|\sum_{k=1}^{j} x_{\sigma(k)} - \frac{j}{2^{N-1}}\sum_{k=1}^{2^{N-1}} x_{\sigma(k)}\right\|$$

$$+\left\|\frac{j}{2^{N-1}}\sum_{k=1}^{2^{N-1}} x_{\sigma(k)} - \frac{j}{2^N}\sum_{k=1}^{2^N} x_k\right\|$$

$$\le \frac{j}{2^{N-1}} C\left(\sum_{k=1}^{2^N}\|x_k\|^p\right)^{1/p} + C\left(\sum_{k=1}^{2^N}\|x_k\|^p\right)^{1/p}\cdot\left(\frac{4}{1-\varepsilon} - \frac{j}{2^{N-2}} - 2\right)$$

$$= C\left(\sum_{k=1}^{2^N}\|x_k\|^p\right)^{1/p}\cdot\left(\frac{4}{1-\varepsilon} - \frac{j}{2^{N-1}} - 2\right).$$

Now consider the case $j > 2^{N-1}$. Using the triangle inequality and then properties (a) and (d), we obtain:

$$\left\|\sum_{k=1}^{j} x_{\sigma(k)} - \frac{j}{2^N}\sum_{k=1}^{2^N} x_{\sigma(k)}\right\| \le \left(2 - \frac{j}{2^{N-1}}\right)\left\|\sum_{k=1}^{2^{N-1}} x_{\sigma(k)} - \frac{1}{2}\sum_{k=1}^{2^N} x_{\sigma(k)}\right\|$$

$$+\left\|\sum_{k=2^{N-1}+1}^{j} x_{\sigma(k)} - \frac{j-2^{N-1}}{2^{N-1}}\sum_{k=2^{N-1}+1}^{2^N} x_{\sigma(k)}\right\|$$

$$\le \left(2 - \frac{j}{2^{N-1}}\right)\cdot C\left(\sum_{k=1}^{2^N}\|x_k\|^p\right)^{1/p}$$

$$+C\left(\sum_{k=2^{N-1}+1}^{2^N}\|x_{\sigma(k)}\|^p\right)^{1/p}\left(\frac{4}{1-\varepsilon} - \frac{j}{2^{N-2}} - 2\right).$$

By property (b), the second term in the last sum is bounded by

$$C\left(\sum_{k=1}^{2^N}\|x_k\|^p\right)^{1/p}\cdot\varepsilon\left(\frac{4}{1-\varepsilon} - \frac{j}{2^{N-2}} - 2\right),$$

and the whole sum is no larger than

$$C\left(\sum_{k=1}^{2^N}\|x_k\|^p\right)^{1/p}\cdot\left(\frac{4\varepsilon}{1-\varepsilon}+2-\frac{j}{2^{N-1}}-\varepsilon\frac{j}{2^{N-2}}-2\varepsilon\right).$$

Recalling that $\varepsilon=(1/2)^{1/p}$ is positive, we see that the coefficient in parentheses is bounded by

$$\frac{4\varepsilon}{1-\varepsilon}+2-\frac{j}{2^{N-1}}=\frac{4}{1-\varepsilon}-\frac{j}{2^{N-1}}-2.$$

This completes the proof of the lemma. □

In what follows instead of (1) we shall use the inequality

$$\left\|\sum_{k=1}^{j}x_{\sigma(k)}-\frac{j}{2^n}\sum_{k=1}^{2^n}x_k\right\|\le C_p\left(\sum_{k=1}^{2^n}\|x_k\|^p\right)^{1/p},$$

where $C_p=\frac{4}{1-\varepsilon}-2$.

THEOREM 7.1.1 (V. M. KADETS [35]). *A space X has infratype p with constant C if and only if it has property* A_1 *with exponent p and some constant K.*

PROOF. We have already proved the implications $\mathrm{A}_1\Rightarrow\mathrm{A}_2\Rightarrow$ infratype $\Rightarrow$ A_2. Here we shall use Lemma 7.1.2 to prove the implication $\mathrm{A}_2\Rightarrow\mathrm{A}_1$.

Consider an arbitrary finite collection $\{x_k\}_{k=1}^m$ of elements of X. Let n be such that $2^n>m$. Introduce auxiliary elements $\{y_k\}_{k=1}^{2^n}$ as follows:

$$y_k=\begin{cases}x_k-\frac{1}{m}\sum_{i=1}^m x_i & \text{if } 1\le k\le m\\ 0 & \text{if } m<k\le 2^n.\end{cases}$$

Obviously, $\sum_{k=1}^{2^n}y_k=0$. Lemma 7.1.2 yields a permutation σ such that

$$\left\|\sum_{k=1}^{j}y_{\sigma(k)}\right\|\le C_p\left(\sum_{k=1}^{2^n}\|y_k\|^p\right)^{1/p}$$

for all $j\le 2^n$. Since $y_k=0$ for $k>m$, we conclude that there exists a permutation π of the set $\{1,\dots,m\}$ such that

$$\left\|\sum_{k=1}^{j}\left(x_{\pi(k)}-\frac{1}{m}\sum_{i=1}^{m}x_i\right)\right\|\le C_p\left(\sum_{k=1}^{m}\|y_k\|^p\right)^{1/p}$$

for all $j \le m$. For this permutation, we have

$$\left\| \sum_{k=1}^{j} x_{\pi(k)} - \frac{j}{m} \sum_{i=1}^{m} x_i \right\| \le C_p \left(\sum_{k=1}^{m} \left(\|x_k\| + \frac{1}{m} \left\| \sum_{i=1}^{m} x_i \right\| \right)^p \right)^{1/p}$$

$$\le C_p \left(\sum_{k=1}^{m} \|x_k\|^p \right)^{1/p} + C_p m^{\frac{1}{p}-1} \sum_{k=1}^{m} \|x_k\|$$

for all $j \le m$. Using the Hölder inequality, we finally obtain

$$\left\| \sum_{k=1}^{j} x_{\pi(k)} - \frac{j}{m} \sum_{k=1}^{m} x_k \right\| \le 2C_p \left(\sum_{k=1}^{m} \|x_k\|^p \right)^{1/p}. \qquad \square$$

Thus, we have shown that the validity of one of the version of Permutation Lemma (condition A_1) is equivalent to the space having an infratype. To prove a full analogue of Steinitz's theorem we also need a Rounding-off-Coefficients Lemma.

LEMMA 7.1.3 (M. I. OSTROVSKII [65]). *Let the Banach space X have infratype p with constant C. Further, let $\{x_i\}_{i=1}^n$ be an arbitrary finite collection of elements of X, $\{\lambda_i\}_{i=1}^n$ be a collection of numerical coefficients, $0 \le \lambda_i \le 1$, and $x = \sum_{i=1}^n \lambda_i x_i$. Then there exists coefficients $\{\theta_i\}_{i=1}^n$, each equal to 0 or 1, such that*

$$\left\| x - \sum_{i=1}^{n} \theta_i x_i \right\| \le C \left(\sum_{i=1}^{n} \|x_i\|^p \right)^{1/p}.$$

PROOF. This is a direct consequence of Lemma 2.3.3. $\square$

We can now state the main result of this section.

THEOREM 7.1.2. *Let the space X have infratype p, let $\sum_{k=1}^\infty x_k$ be a series in X, and let $s = \sum_{k=1}^\infty x_k$. If the terms of the series satisfy the condition $\sum_{k=1}^\infty \|x_k\|^p < \infty$, then $\mathrm{SR}(\sum_{k=1}^\infty x_k)$ is the affine subspace $s + \Gamma^\perp$, where $\Gamma^\perp \subset X$ is the annihilator of the set $\Gamma \subset X^*$ of convergence functionals.* $\square$

EXERCISE 7.1.1. Show that if the space X does not have infratype p, then the assertion of Lemma 7.1.3 is false.

EXERCISE 7.1.2. Suppose that in the space X the series $\sum_{k=1}^\infty \|x_k\|^p$ is divergent whenever the series $\sum_{k=1}^\infty x_k$ is perfectly divergent. Show that X has infratype p.

CONJECTURE. If a space X has infratype p, then for any perfectly divergent series $\sum_{k=1}^\infty x_k$ the series $\sum_{k=1}^\infty \|x_k\|^p$ diverges.

§2. A Technique for Transferring Examples with Nonlinear Sum Range to Arbitrary Infinite-Dimensional Banach Spaces

In this section we shall prove a theorem of V. M. Kadets on the existence of series with nonlinear sum range in any infinite-dimensional Banach space. The auxiliary results needed for this theorem will be stated in general form, to facilitate their use in the next section. The first of these results is based on Lemma 6.3.4.

THEOREM 7.2.1. *Let X and Y be Banach spaces and suppose that $X \overset{\mathrm{f}}{\Rightarrow} Y$. Then there exists a constant D, $0 \le D \le 12$, such that, for each basic sequence $\{e_k\}_{k=1}^{\infty}$ in X and each sequence $\{A_k\}_{k=1}^{\infty}$ of positive numbers that tends monotonically to infinity one can find a basic sequence $\{g_k\}_{k=1}^{\infty}$ in Y such that*

$$\left\|\sum_{k=1}^{n} t_k e_k\right\| \le \left\|\sum_{k=1}^{n} t_k g_k\right\| \le (D + A_n)\left\|\sum_{k=1}^{n} t_k e_k\right\| \tag{1}$$

for all n and all collections of numbers $\{t_k\}_{k=1}^{n}$.

PROOF. Let C be the basic constant of the sequence $\{e_k\}_{k=1}^{\infty}$. Choose integers $0 = n_0 < n_1 < n_2 < \dots$ such that $A_{n_k} \ge 2^{k+7}/C$ for $k = 1, 2, \dots$. Denote

$$E_j = \mathrm{Lin}\{e_{n_j+1}, \dots, e_{n_{j+1}}\}, \quad j = 0, 1, 2, \dots .$$

Using Theorem 6.3.4, pick a basic sequence $\{Y_j\}_{j=0}^{\infty}$ of subspaces of Y with basic constant $1-\varepsilon$, such that $d(E_j, Y_j) \le 1+\varepsilon$, $j = 0, 1, 2, \dots$. Thanks to this inequality, one can find in each subspace Y_j a basis $\{y_i\}_{i=n_j+1}^{n_{j+1}}$ with the property that the double inequality

$$\left\|\sum_{i=n_j+1}^{n_{j+1}} t_i e_i\right\| \le \left\|\sum_{i=n_j+1}^{n_{j+1}} t_i y_i\right\| \le (1+\varepsilon)\left\|\sum_{k=n_j+1}^{n_{j+1}} t_i e_i\right\| \tag{2}$$

holds for any choice of numbers t_i.

Define the sought-for sequence $\{g_i\}_{i=1}^{\infty}$ as follows: $g_i = 2^{j+3} y_i$ for $i \in \{n_j + 1, \dots, n_{j+1}\}$, $j = 0, 1, 2, \dots$. Put $D = 8(1+\varepsilon)$. Then for $n \le n_1$ inequality (1) follows from (2). Now let us prove (1) for $n > n_1$. Let j be such that $n_j < n \le n_{j+1}$, and let $\{t_k\}_{k=1}^{n}$ be arbitrary numbers. For the sake of convenience, put $t_k = 0$ for $k > n$. Using the definition of the elements g_j and inequality (2), we obtain

$$\left\|\sum_{k=1}^{n} t_k g_k\right\| \le \sum_{i=0}^{j}\left\|\sum_{k=n_i+1}^{n_{i+1}} t_k g_k\right\| \le \sum_{i=0}^{j}\left[(1+\varepsilon)2^{i+3}\left\|\sum_{k=n_i+1}^{n_{i+1}} t_k e_k\right\|\right]$$

$$\le (1+\varepsilon)2^{j+4} \max_i \left\{\left\|\sum_{k=n_i+1}^{n_{i+1}} t_k e_k\right\|\right\}.$$

By Remark 6.3.1, the last quantity is estimated from above as follows:

$$\frac{2}{C}(1+\varepsilon)2^{j+4}\left\|\sum_{k=1}^{n} t_k e_k\right\| \le A_{n_j}\left\|\sum_{k=1}^{n} t_k e_k\right\| \le A_n\left\|\sum_{k=1}^{n} t_k e_k\right\|.$$

Thus, one part of inequality (1) is proved. Let us prove the second part. By construction, $\{Y_j\}_{j=0}^{\infty}$ is a basic sequence of subspaces with basic constant $1-\varepsilon > 1/2$. Using Remark 6.3.1, the definition of g_j, and inequality (2), we obtain

$$\left\|\sum_{k=1}^{n} t_k g_k\right\| \ge \frac{1}{4}\max_i\left\{\left\|\sum_{k=n_i+1}^{n_{i+1}} t_k g_k\right\|\right\}$$

$$\ge \frac{1}{4}\max_i\left\{2^{i+3}\left\|\sum_{k=n_i+1}^{n_{i+1}} t_k g_k\right\|\right\} = 2\max_i\left\{2^{i}\left\|\sum_{k=n_i+1}^{n_{i+1}} t_k g_k\right\|\right\}.$$

If we now put $a_i = 2^i\|\sum_{k=n_i+1}^{n_{i+1}} t_k e_k\|$ and $b_i = 2^{-i}$ in the inequality

$$\sum_{i=0}^{\infty}|a_i|\cdot|b_i| \le \max_i|a_i|\cdot\sum_{k=0}^{\infty}|b_k|,$$

the preceding inequality yields

$$\left\|\sum_{k=1}^{n} t_k g_k\right\| \ge \sum_{i=0}^{\infty}\left\|\sum_{k=n_i+1}^{n_{i+1}} t_k e_k\right\| \ge \left\|\sum_{k=1}^{n} t_k e_k\right\|,$$

which completes the proof of the theorem. □

LEMMA 7.2.1. *Let X be a Banach space with a basis $\{e_k\}_{k=1}^{\infty}$ and let $\{x_k\}_{k=1}^{\infty} \subset X$ be an arbitrary countable set. Then one can find a basis $\{\overline{e}_k\}_{k=1}^{\infty}$ in X such that $\{x_i\}_{i=1}^{\infty} \subset \operatorname{Lin}\{\overline{e}_i\}_{i=1}^{\infty}$, or, in other words, the decomposition of each element x_i in the basis $\{\overline{e}_k\}_{k=1}^{\infty}$ is a finite sum.*

PROOF. We shall use the Krein-Mil'man-Rutman theorem (Theorem 6.3.2). Let C be the basic constant of the basis $\{e_k\}_{k=1}^{\infty}$. With no loss of generality we may assume that $\|e_i\| = 1$, $i = 1, 2, \ldots$. Let $x_i \ne 0$ for all i. Write the decomposition of x_1 in the given basis: $x_1 = \sum_{i=1}^{\infty} a_i e_i$. Let i_1 be an index such that $a_{i_1} \ne 0$, and let $n_1 > i_1$ be a number for which

$$\left\|\sum_{i=n_1+1}^{\infty} a_i e_i\right\| \le C|a_{i_1}|2^{-i_1-3}.$$

Define $\overline{e}_i$ for $i \le n_1$ as follows: $\overline{e}_i = e_i$ for $i \ne i_1$, and

$$\overline{e}_{i_1} = e_{i_1} + \frac{1}{a_{i_1}}\sum_{i=n_1+1}^{\infty} a_i e_i.$$

Then $\|e_i - \overline{e}_i\| \leq C2^{-i-3}$, $i = 1, \ldots, n_1$, and, by the Krein-Mil'man-Rutman theorem, the system $\overline{e}_1, \ldots, \overline{e}_{n_1}, e_{n_1+1}, e_{n_1+2}, \ldots$ is a basis in X. Moreover, $x_1 = \sum_{i=1}^{n_1} a_i \overline{e}_i$.

We will now proceed by induction. Specifically, let us show that one can choose a sequence of natural numbers $n_1 < n_2 < \ldots$ and a system of vectors $\{\overline{e}_i\}_{i=1}^{\infty} \subset X$ such that, for any $i \in \mathbf{N}$,

$$\|e_i - \overline{e}_i\| \leq C2^{-i-3}, \quad x_i \in \mathrm{Lin}\{\overline{e}_1, \ldots, \overline{e}_{n_i}\}. \tag{3}$$

Thus, suppose that we already found numbers $n_1, \ldots, n_i$ and vectors $\overline{e}_1, \ldots, \overline{e}_{n_i}$ such that relations (3) hold. Then, by the Krein-Mil'man-Rutman theorem, the system

$$\overline{e}_1, \ldots, \overline{e}_{n_i}, e_{n_i+1}, e_{n_i+2}, \ldots$$

is a basis in X. Let us decompose the vector x_{i+1} with respect to this basis:

$$x_{i+1} = \sum_{k=1}^{n_i} b_k \overline{e}_k + \sum_{k=n_i+1}^{\infty} b_k e_k.$$

If only finitely many of the numbers b_k there are different from zero, we can take for n_{i+1} a natural number such that $b_k = 0$ for $k > n_{i+1}$, and take the vectors $\overline{e}_k$ with $n_i < k \leq n_{i+1}$ equal to the corresponding vectors e_k. Hence, we need to examine only the case when infinitely many b_k's are different from zero. Let $r > n_i$ be an index such that $b_r \neq 0$. Take for n_{i+1} a natural number such that $n_{i+1} > r$ and

$$\left\| \sum_{k=n_{i+1}+1}^{\infty} b_k e_k \right\| \leq C b_r 2^{-r-3}.$$

Now define $\overline{e}_k$ for $n_i < k \leq n_{i+1}$ as follows: $\overline{e}_k = e_k$ for $k \neq r$, and

$$\overline{e}_r = e_r + \frac{1}{b_r} \sum_{k=n_{i+1}+1}^{\infty} b_k e_k.$$

Then $x_{i+1} = \sum_{k=1}^{n_{i+1}} b_k \overline{e}_k$ and $\|e_k - \overline{e}_k\| \leq C2^{-k-3}$ for $k \leq n_{i+1}$. Continuing this induction proces, we obtain a system $\{\overline{e}_k\}_{k=1}^{\infty}$. By the Krein-Mil'man-Rutman theorem, this system is a basis in X and $x_i \in \mathrm{Lin}\{\overline{e}_k\}$ for all i. □

THEOREM 7.2.2. *Let X and Y be Banach spaces, $X \overset{\mathrm{f}}{\Rightarrow} Y$. Suppose that X has a basis $\{e_k\}_{k=1}^{\infty}$ and let $\sum_{k=1}^{\infty} x_k$ be a series in X such that $\mathrm{SR}(\sum_{k=1}^{\infty} x_k)$ is not a linear set. Then for any monotone sequence of positive numbers $\{a_k\}_{k=1}^{\infty}$ with $a_k \to \infty$, $k \to \infty$, there exists a series $\sum_{k=1}^{\infty} y_k$ in Y such that $\mathrm{SR}(\sum_{k=1}^{\infty} y_k)$ is not a linear set and $\|y_k\| \leq a_k \|x_k\|$ for all $k \in \mathbf{N}$.*

PROOF. To simplify the arguments we will assume that $a_1 > 12$. This can be achieved by multiplying all numbers a_i by a sufficiently large number K and dividing all the vectors x_i by K. Since $\mathrm{SR}(\sum_{k=1}^{\infty} x_k)$ is not a linear set, there exist vectors $u, v \in \mathrm{SR}(\sum_{k=1}^{\infty} x_k)$ and a real number λ such that

$$\lambda u + (1-\lambda)v \notin \mathrm{SR}(\sum_{k=1}^{\infty} x_k).$$

With no loss of generality we may assume that all the vectors x_i, u, and v belong to $\mathrm{Lin}\{e_k\}_{k=1}^{\infty}$ (by Lemma 7.2.1, this can be ensured by a small perturbation of the initial basis). Let π and σ be permutations such that

$$\sum_{k=1}^{\infty} x_{\pi(k)} = u, \quad \sum_{k=1}^{\infty} x_{\sigma(k)} = v.$$

Denote

$$b_j = \max\left\{\left\|u - \sum_{k=1}^{j} x_{\pi(k)}\right\|, \left\|v - \sum_{k=1}^{j} x_{\sigma(k)}\right\|\right\}.$$

Clearly, $\lim_{j\to\infty} b_j = 0$. Pick numbers $n_1 < n_2 < \dots$, sufficiently large so that for each j the elements

$$u - \sum_{k=1}^{j} x_{\pi(k)}, \quad v - \sum_{k=1}^{j} x_{\sigma(k)}, \quad x_j$$

belong to the linear span of the corresponding set $\{e_1, \dots, e_{n_j}\}$. Let $\{A_k\}_{k=1}^{\infty}$ be a monotone sequence of positive numbers such that $A_k \to \infty$,

$$12 + A_{n_j} \le a_j, \quad \lim_{j\to\infty} b_j A_{n_j} = 0.$$

Now let us use Theorem 7.2.1 and select a set of vectors $\{g_k\}_{k=1}^{\infty} \subset Y$ such that, for any collection of numbers t_k,

$$\left\|\sum_{k=1}^{n} t_k e_k\right\| \le \left\|\sum_{k=1}^{n} t_k g_k\right\| \le (12 + A_n)\left\|\sum_{k=1}^{n} t_k e_k\right\|.$$

The linear operator $T\colon \mathrm{Lin}\{g_k\}_{k=1}^{\infty} \to \mathrm{Lin}\{e_k\}_{k=1}^{\infty}$ acting as $Tg_k = e_k$, $k \in \mathbf{N}$, is continuous and injective. Now define the terms y_k of the sought-for series by $y_k = T^{-1}x_k$. Also, denote $\overline{u} = T^{-1}u$, $\overline{v} = T^{-1}v$. Since $x_k \in \mathrm{Lin}\{e_k\}_{k=1}^{n_k}$, and hence $y_k \in \mathrm{Lin}\{g_k\}_{k=1}^{n_k}$, we have $\|y_k\| \le (12 + A_{n_k})\|x_k\| \le a_k\|x_k\|$. In a similar manner we can show that

$$\max\left\{\left\|\overline{u} - \sum_{k=1}^{j} y_{\pi(k)}\right\|, \left\|\overline{v} - \sum_{k=1}^{j} y_{\sigma(k)}\right\|\right\} \le b_j(12 + A_{n_j}),$$

and since the right-hand side of this inequality tends to zero, the vectors $\overline{u}$ and $\overline{v}$ belong to $\mathrm{SR}(\sum_{k=1}^{\infty} y_k)$. We claim that $\lambda\overline{u} + (1-\lambda)\overline{v} \notin \mathrm{SR}(\sum_{k=1}^{\infty} y_k)$. Indeed, suppose that the opposite holds, i.e., for some permutation ν the series $\sum_{k=1}^{\infty} y_{\nu(k)}$ converges to $\lambda\overline{u} + (1-\lambda)\overline{v}$. Thanks to the continuity of T,

$$\sum_{k=1}^{\infty} x_{\nu(k)} = \sum_{k=1}^{\infty} T y_{\nu(k)} = T(\lambda\overline{u} + (1-\lambda)\overline{v}) = \lambda u + (1-\lambda)v$$

which contradicts the assumption that $\lambda u + (1-\lambda)v \notin \mathrm{SR}(\sum_{k=1}^{\infty} x_k)$. Therefore, the constructed series $\sum_{k=1}^{\infty} y_k$ satisfies all the requirements of the theorem. □

COROLLARY 7.2.1. *In any infinite-dimensional Banach space X there are series with a nonlinear sum range.*

PROOF. By Dvoretzky's theorem, l_2 is finitely representable in any infinite-dimensional space. Consequently, $L_2[0,1] \overset{\mathrm{f}}{\Rightarrow} X$. Since in $L_2[0,1]$ there are series with nonlinear sum range, it follows that such series exists in X as well. □

COROLLARY 7.2.2. *In any infinite-dimensional Banach space there are series whose sum range consists of two points.*

PROOF. Notice that if in Theorem 7.2.2 $\mathrm{SR}(\sum_{k=1}^{\infty} x_k)$ consists of two points, then the same is true about $\mathrm{SR}(\sum_{k=1}^{\infty} y_k)$. Indeed, it is readily seen that the operator T constructed in the proof of the theorem extends to an injective operator $T\colon \overline{\mathrm{Lin}}\{g_k\}_{k=1}^{\infty} \to X$. Since $\mathrm{SR}(\sum_{k=1}^{\infty} x_k) \supset T(\mathrm{SR}(\sum_{k=1}^{\infty} y_k))$, we see that $\mathrm{SR}(\sum_{k=1}^{\infty} y_k)$ cannot contain points different from $\overline{u}$ and $\overline{v}$. To prove the corollary it suffices to use Dvoretzky's theorem and the existence in L_2 of a series with a two-point sum range. □

EXERCISE 7.2.1. In Theorem 7.2.1 we omitted to show that $\{g_k\}_{k=1}^{\infty}$ is a basic sequence (we did not use that property further). Complete the omitted part of the proof and estimate the basic constant of $\{g_k\}_{k=1}^{\infty}$.

EXERCISE 7.2.2. Under the assumptions of the Krein-Mil'man-Rutman theorem, show that if $\{e_i\}_{i=1}^{\infty}$ is a basis in X, then $\{y_i\}_{i=1}^{\infty}$ is also a basis (we did use this fact, without explicit mention, in the proof of Lemma 7.2.1).

EXERCISE 7.2.3. Show that for $p > 2$ there is no infinite-dimensional Banach space in which the condition $\sum_{k=1}^{\infty} \|x_k\|^p < \infty$ is sufficient for the linearity of the set $\mathrm{SR}(\sum_{k=1}^{\infty} x_k)$.

EXERCISE 7.2.4. Find a more precise estimate of the constant D appearing in Theorem 7.2.1.

§3. Series in Spaces That Are Not B-Convex

In this section we will show that results analogous to those obtained in §1 do not hold in spaces that are not B-convex.

DEFINITION 7.3.1. A series $\sum_{k=1}^{\infty} x_k$ of elements of a Banach space is said to be *dominated by a numerical series* $\sum_{k=1}^{\infty} a_k$ if $\|x_k\| \leq a_k$ for all $k \in \mathbf{N}$.

DEFINITION 7.3.2. We say that *Steinitz's theorem holds in the space X relative to the sequence $\{a_k\}_{k=1}^{\infty}$ of positive numbers*, and we write $X \in \mathcal{S}(\{a_k\}_{k=1}^{\infty})$, if any series in X that is dominated by the series $\sum_{k=1}^{\infty} a_k$ has a linear sum range.

Needless to say, if $\sum_{k=1}^{\infty} a_k < \infty$, then Steinitz's theorem relative to the sequence $\{a_k\}_{k=1}^{\infty}$ holds in any Banach space. Our goal is to prove that if $\sum_{k=1}^{\infty} a_k$ diverges and X is not a B-convex space, then $X \notin \mathcal{S}(\{a_k\}_{k=1}^{\infty})$. We will start by proving this assertion for the case $X = L_1[0,1]$, which was dealt with by P. A. Kornilov [46] in a somewhat different formulation.

THEOREM 7.3.1. *Suppose that $\{a_k\}_{k=1}^{\infty}$ is a monotonically nonincreasing sequence of positive numbers and $\sum_{k=1}^{\infty} a_k = \infty$. Then in $L_1[0,1]$ there exist a series $\sum_{k=1}^{\infty} x_k$ dominated by $\sum_{k=1}^{\infty} a_k$, such that $\mathrm{SR}(\sum_{k=1}^{\infty} x_k)$ is not a linear set.*

PROOF. Pick positive numbers $\{b_k\}_{k=1}^{\infty}$ such that $b_k \leq \min\{a_{2k}, a_{2k-1}\}$, $\lim_{k\to\infty} b_k = 0$, and $\sum_{k=1}^{\infty} b_k = \infty$. Denote the segment $[0,1]$ by I_0. Choose a sequence of segments I_k, $k \geq 1$, and numbers n_k, $0 = n_1 < n_2 < \ldots$, such that $|I_k| < b_k$, the segments $\{I_k\}_{k=n_j+1}^{n_{j+1}}$ are pairwise disjoint, and for j one has that $I_{j-1} = \bigcup_{k=n_j+1}^{n_{j+1}} I_k$. This selection is carried out by means of the following inductive process: first partition the segment I_0 into a finite number of disjoint segments I_k, then partition each I_k into a finite number of disjoint segments, and so on. The condition $\sum_{k=1}^{\infty} b_k = \infty$ guarantees that one can satisfy the inequalities $|I_k| \leq b_k$, $k = 1, 2, \ldots$. Now define the sought-for series $\sum_{k=1}^{\infty} x_k$ as follows:

$$x_{2j}(t) = \begin{cases} 1 & \text{if } t \in I_j \\ 0 & \text{if } t \notin I_j, \end{cases}$$

and

$$x_{2j-1} = -x_{2j}.$$

Since $\|x_{2j}\| = \|x_{2j-1}\| = |I_j| < b_j$, we see that $\lim_{k\to\infty} \|x_k\| = 0$. The inequality $b_j < \min\{a_{2j}, a_{2j-1}\}$ shows that the series $\sum_{k=1}^{\infty} x_k$ is dominated by the series $\sum_{k=1}^{\infty} a_k$. It is clear that $\sum_{k=1}^{\infty} x_k = 0$ and that, by analogy with Marcinkiewicz's example (Example 3.1.2), one can rearrange the terms of our series so that it will converge to the function identically equal to 1. At the same time, the terms of the constructed series are integer-valued functions, and consequently the series cannot be rearranged so that it will converge to the function identically equal to $1/2$. □

LEMMA 7.3.1. *Suppose that the space X is not B-convex. Then X has no finite-codimensional B-convex subspaces.*

PROOF. It suffices to prove the lemma for subspaces of codimension 1. We shall argue by reductio ad absurdum. So, let Y be a codimension-1 subspace of X which is a B-convex space. Then Y has some type p, $1 < p \leq 2$, with constant C.

We claim than in this case X also has type p. Indeed, let $f \in X^*$ be a functional such that $\operatorname{Ker} f = Y$ and $\|f\| = 1$, and let $e \in S(X)$ be such that $f(e) > 1/2$. Further, let $\{x_i\}_{i=1}^n$ be an arbitrary finite collection of elements of X. Write each x_i as $x_i = a_i e + y_i$, where $a_i \in \mathbf{R}$, $y_i \in Y$. Then

$$|a_i| \leq 2|(f(a_i e)| = 2|f(a_i e + y_i)| \leq 2\|x_i\|$$

and $\|y_i\| \leq \|x_i\| + \|a_i e\| \leq 3\|x_i\|$. Using the definition of the type and Khinchin's inequality, we obtain

$$\mathbf{E}\left\{\left\|\sum_{i=1}^n r_i x_i\right\|\right\} \leq \mathbf{E}\left\{\left\|\sum_{i=1}^n r_i y_i\right\|\right\} + \mathbf{E}\left\{\left|\sum_{i=1}^n r_i a_i\right|\right\}$$

$$\leq C\left(\sum_{i=1}^n \|y_i\|^p\right)^{1/p} + K\left(\sum_{i=1}^n |a_i|^2\right)^{1/2}$$

$$\leq 3C\left(\sum_{i=1}^n \|x_i\|^p\right)^{1/p} + K\left(\sum_{i=1}^n |a_i|^p\right)^{1/p} \leq (3C + 2K)\left(\sum_{i=1}^n \|x_i\|^p\right)^{1/p}.$$

Therefore, X has type $p > 1$, which contradicts our hypothesis. □

Finally, let us prove the main result of this section.

THEOREM 7.3.2 (V. M. KADETS). *Suppose that the space Y is not B-convex and let $\{a_k\}_{k=1}^\infty$ be a sequence of positive numbers such that $\sum_{k=1}^\infty a_k = \infty$. Then $Y \notin \mathcal{S}(\{a_k\}_{k=1}^\infty)$.*

PROOF. We need to exhibit a series $\sum_{k=1}^\infty y_k$ in Y, dominated by the series $\sum_{k=1}^\infty a_k$, for which the set $\mathrm{SR}(\sum_{k=1}^\infty y_k)$ is not linear. Since rearranging the terms of a series does not change its sum range, and since by making the numbers a_i smaller we would only make our problem more difficult, it suffices to examine the case when the sequence $\{a_k\}_{k=1}^\infty$ converges monotonically to zero.

Choose two sequences of positive numbers $\{b_k\}$ and $\{c_k\}$ such that $a_k = b_k c_k$, $b_k \uparrow \infty$, $c_k \downarrow 0$, and $\sum_{k=1}^\infty c_k = \infty$. By Theorem 7.3.1, in the space $L_1[0,1]$ one can exhibit a series $\sum_{k=1}^\infty x_k$, dominated by the series $\sum_{k=1}^\infty c_k$, such that the set $\mathrm{SR}(\sum_{k=1}^\infty x_k)$ is not linear. By Proposition 5.1.1, $L_1[0,1] \xrightarrow{\mathrm{f}} l_1$. Hence, by Lemma 7.3.1, $L_1[0,1] \overset{\mathrm{f}}{\Rightarrow} Y$. Now using Theorem 7.2.2 we see that in Y there exists a series $\sum_{k=1}^\infty y_k$ with nonlinear sum range, dominated by the series $\sum_{k=1}^\infty b_k \|x_k\|$. Since $b_k c_k = a_k$, the series $\sum_{k=1}^\infty y_k$ is dominated by the series $\sum_{k=1}^\infty a_k$. □

Combining this result with the results of §1, we conclude the the analogues of Steinitz's theorem considered here hold in the B-convex spaces and only in them.

EXERCISE 7.3.1. Prove Lemma 7.3.1 without using the notions of type and infratype.

EXERCISE 7.3.2. Let $p \in [1,2]$, $a_k > 0$ for all $k \in \mathbf{N}$, and $\sum_{k=1}^{\infty} a_k^p = \infty$. Show that under these assumptions $L_p[0,1] \notin \mathcal{S}(\{a_k\}_{k=1}^{\infty})$.

EXERCISE 7.3.3. Suppose the space X does not have infratype p and $\sum_{k=1}^{\infty} a_k^p = \infty$. Show that $X \notin \mathcal{S}(\{a_k\}_{k=1}^{\infty})$.

We do not know the answers to the following problems:

PROBLEM 7.3.1. Suppose the space X is not B-convex and $\{a_k\}_{k=1}^{\infty}$ is a sequence such that $a_k \downarrow 0$ and $\sum_{k=1}^{\infty} a_k = \infty$. Does there exist a series $\sum_{k=1}^{\infty} x_k$ in X with nonlinear sum range and such that $\|x_k\| = a_k$ for all k?

PROBLEM 7.3.2. Can Theorem 7.1.2 hold in spaces X that do not have infratype p?

CHAPTER 8
REARRANGEMENTS OF SERIES IN TOPOLOGICAL VECTOR SPACES

There are many types of convergence that cannot be defined in terms of a norm. In this chapter we will examine the problem of the validity of Steinitz's theorem when the convergence of series is understood in the sense of the weak topology of a normed space (§1), or when one deals with convergence in measure for series of functions (§2), and finally when one deals with convergence of series in a nuclear Fréchet space (§3). Steinitz-type problems in topological vector spaces are relatively poorly investigated, and we collected here results that are the most definitive in nature. In the first two sections we construct a number of counterexamples, and in the third section we give a result of W. Banaszczyk which extends Steinitz's theorem from finite-dimensional spaces to nuclear Fréchet spaces.

§1. Weak and Strong Sum Range

Let $\sum_{k=1}^{\infty} x_k$ be s series of elements of a Banach space X. By analogy with the notion of the sum range, we introduce the notion of the *weak sum range* as follows: $\mathrm{WSR}(\sum_{k=1}^{\infty} x_k)$ is the set of elements to which a rearrangement of the series $\sum_{k=1}^{\infty} x_k$ converges in the weak topology.

REMARK 8.1.1. Let $\sum_{k=1}^{\infty} x_k$ be a series of elements in a Banach space X, let $\Gamma \subset X^*$ be the set of convergence functionals of this series, and let $a \in \mathrm{WSR}(\sum_{k=1}^{\infty} x_k)$. Then the following obvious inclusions hold:

$$\mathrm{SR}(\sum_{k=1}^{\infty} x_k) \subset \mathrm{WSR}(\sum_{k=1}^{\infty} x_k) \subset a + \Gamma^{\perp}.$$

Hence, if Steinitz's theorem holds for the sum range, i.e., $\mathrm{SR}(\sum_{k=1}^{\infty} x_k) = a + \Gamma^{\perp}$, then it also holds for the weak sum range.

The weak topology is much closer in its properties to a topology of finite-dimensional space than the norm topology. This is particularly evident in reflexive spaces, where from any bounded sequence one can extract a weakly convergent subsequence. Nevertheless, the weak sum range of a series can fail to be a linear set even in such a nice space as Hilbert space. A counterexample was provided by V. M. Kadets [41]; its construction is technically quite difficult. Before we turn to the exposition of that construction, let us point out that in Marcinkiewicz's example the weak sum range is linear (the reader is invited to check this fact by himself).

To facilitate the exposition, let us introduce a number of auxiliary definitions. An ordered set A with an order relation F will be denoted by (A, F). The relation "a_1 succeeds a_2 in the order F" will be written as $a_1 >_F a_2$. We say that the element a_1 is a *direct successor* of the element a_2 if $a_1 \neq a_2$, $a_1 >_F a_2$, and the relations $a_1 >_F a$, $a >_F a_2$ imply that either $a = a_1$ or $a = a_2$. The set of direct successors of the element a in the order F will be denoted by $F(a)$. If $B \subset A$, then we put $F(B) = \bigcup_{b \in B} F(b)$. We define the set $F^n(a)$ as follows:

$$F^1(a) = F(a),\ F^2(a) = F(F(a)),\ \ldots,\ F^n(a) = F(F^{n-1}(a)).$$

DEFINITION 8.1.1. An order relation F on the set $\mathbf{N}$ of natural numbers is said to be *tree-like* if:

a) 1 is the smallest element in $(\mathbf{N}, F)$;

b) the set $F(n)$ is finite for any $n \in \mathbf{N}$;

c) $F(n) \cap F(m) = \emptyset$ for $n \neq m$;

d) $\bigcup_{k=1}^{\infty} F^k(1) = \mathbf{N} \setminus \{1\}$.

DEFINITION 8.1.2. A sequence $\{x_n\}_{n=1}^{\infty}$ of elements of a Hilbert space is said to be *branched* with respect to the tree-like order relation F (or simply F-*branched*) if the following conditions are satified:

1) $x_n = \sum_{i \in F(n)} x_i$ for any $n \in \mathbf{N}$;

2) $\{x_i\}_{i \in F^n(1)}$ is a linearly independent set of vectors for any $n \in \mathbf{N}$.

Let x be an arbitrary element of the linear span of an F-branched sequence $\{x_n\}_{n=1}^{\infty}$. By condition 1), for any sufficiently large $n \in \mathbf{N}$ one can represent x in the form $x = \sum_{i \in F^n(1)} t_i x_i$. By condition 2), for fixed n such a representation is unique. Let $n > m$, $k \in F^m(1)$, and

$$x = \sum_{i \in F^n(1)} t_i x_i. \tag{1}$$

Put

$$T_k(x) = \sum_{i >_F k,\ i \in F^n(1)} t_i x_i.$$

Using the uniqueness of representation (1) for each fixed n and condition 1) of Definition 8.1.2, which allows us to relate the representations (1) for different values of n, we deduce that $T_k(x)$ does not depend on n and in this way a linear operator T_k (possibly unbounded) is defined on the linear span of the sequence $\{x_n\}_{n=1}^{\infty}$. Let us list a number of elementary properties of the operators T_k.

LEMMA 8.1.1. a) *If x is a linear combination with integer coefficients of elements of the F-branched sequence $\{x_n\}_{n=1}^{\infty}$, then for any n all coefficients t_i in representation* (1) *and in the definition of $T_k(x)$ are also integers.*

b) *The identity $\sum_{i\in F^n(1)} T_i(x) = x$ holds for any $n \in \mathbf{N}$ and any $x \in \mathrm{Lin}\{x_n\}_{n=1}^{\infty}$.* □

DEFINITION 8.1.3. A system $\{x_n^*\}_{n=1}^{\infty}$ of elements of a Hilbert space is said to be *conjugate* to the F-branched sequence $\{x_n\}_{n=1}^{\infty}$ if

a) for any $n \in \mathbf{N}$ and any $i, j \in F^n(1)$ one has that $\langle x_i, x_j^*\rangle = \delta_{ij}$, where δ_{ij} is the Kronecker symbol;

b) $\mathrm{Lin}\{x_i^*\}_{i\in F^n(1)} = \mathrm{Lin}\{x_i\}_{i\in F^n(1)}$, $n = 1, 2, \dots$.

LEMMA 8.1.2. *For any F-branched sequence $\{x_n\}_{n=1}^{\infty}$ there exists a unique conjugate system $\{x_n^*\}_{n=1}^{\infty}$. The operator P_n acting as $P_n(x) = \sum_{i\in F^n(1)}\langle x, x_i^*\rangle\, x_i$ is the orthogonal projection of the ambient space onto $\mathrm{Lin}\{x_i\}_{i\in F^n(1)}$.*

PROOF. The existence of a conjugate system follows from the linear independence of the set $\{x_i\}_{i\in F^n(1)}$, while its uniqueness follows from condition b) of Definition 8.1.3. Further, the fact that P_n is the projection onto $\mathrm{Lin}\{x_i\}_{i\in F^n(1)}$ is a consequence of condition a) of Definition 8.1.3, and the orthogonality of P_n is a consequence of condition b) of the same definition. □

In what follows the notation P_n will be reserved for the projections defined in the preceding lemma.

DEFINITION 8.1.4. An F-branched sequence $\{x_i\}_{i=1}^{\infty}$ is said to be *F-widely spread* if the following conditions are satisfied:

1) for any $n \in \mathbf{N}$, the angle between any two elements of the set $\{x_i\}_{i\in F^n(1)}$ is obtuse;

2) if $i >_F j$, then $\langle x_j^*, x_i\rangle > 0$;

3) if $j \in F^n(1)$ and $i >_F j$, then $P_n(x_i)$ is collinear with x_j;

4) if $j_1, j_2 \in F^n(1)$, $i_1 >_F j_1$, $i_2 >_F j_2$, and $j_1 \neq j_2$, then the vectors $x_{i_1} - P_n(x_{i_1})$ and $x_{i_2} - P_n(x_{i_2})$ form a right angle.

LEMMA 8.1.3. *Suppose that $\{x_i\}_{i=1}^{\infty}$ is an F-widely spread sequence and let $x = \sum_{i\in F^m(1)} x_i t_i$, $n < m$ and $k \in F^n(1)$. Then $\|T_k(x)\| \geq \|x_k\| \cdot \min|t_i|$.*

PROOF. From condition 1) of Definition 8.1.4 it follows that

$$\|T_k(x)\| = \left\| \sum_{i>_F k,\, i\in F^m(1)} t_i x_i \right\| \geq \left\| \sum_{i>_F k,\, i\in F^m(1)} |t_i| x_i \right\|.$$

Using conditions 3) and 2) of Definition 8.1.4 we obtain

$$\|T_k(x)\| \geq \left\| P_n\left(\sum_{i>_F k,\, i\in F^m(1)} |t_i| x_i \right) \right\| = \left\langle x_k^*, \sum_{i>_F k,\, i\in F^m(1)} |t_i| x_i \right\rangle \cdot \|x_k\|$$

$$\geq \min |t_i| \cdot \left\langle x_k^*, \sum_{i >_F k,\, i \in F^m(1)} x_i \right\rangle \cdot \|x_k\| = \|x_k\| \cdot \min |t_i| \langle x_k^*, x_k \rangle = \|x_k\| \cdot \min |t_i|.$$

(In the last [resp. last but one] step we used condition a) [resp. condition 1)] of Definition 8.1.3 [resp. Definition 8.1.2].) □

LEMMA 8.1.4. *Let X denote the linear span of an F-widely spread sequence $\{x_i\}_{i=1}^\infty$. Let $\{d_k\}_{k=1}^\infty$ be a sequence of elements of X that converges weakly to zero. Then, for any $n \in \mathbf{N}$,*

$$\lim_{k \to \infty} \left(\|d_k\|^2 - \sum_{i \in F^n(1)} \|T_i(d_k)\|^2 \right) = 0.$$

PROOF. Consider the equality $d_k = \sum_{i \in F^n(1)} T_i(d_k)$, take its square, and use condition 4) of Definition 8.1.4 to transform the result as follows:

$$\|d_k\|^2 = \sum_{i \in F^n(1)} \|T_i(d_k)\|^2 + \sum_{i,j \in F^n(1),\, i \neq j} \langle T_i(d_k), T_j(d_k) \rangle$$

$$= \sum_{i \in F^n(1)} \|T_i(d_k)\|^2 + \sum_{i,j \in F^n(1),\, i \neq j} \langle P_n T_i(d_k), P_n T_j(d_k) \rangle. \tag{2}$$

By the definition of the projection P_n and condition 3) of Definition 8.1.4, $P_n T_i(d_k) = \langle x_i^*, T_i(d_k) \rangle x_i$, $i \in F^n(1)$; further, by the definition of the operator T_i and condition 4) of Definition 8.1.4, $P_n T_i(d_k) = \langle x_i^*, d_k \rangle x_i$, $i \in F^n(1)$. Since the sequence $\{d_k\}_{k=1}^\infty$ converges weakly to zero,

$$\lim_{k \to \infty} P_n T_i(d_k) = \lim_{k \to \infty} \langle x_j^*, d_k \rangle x_i = 0,$$

which in conjunction with relation (2) completes the proof of the lemma. □

LEMMA 8.1.5. *Let F be a tree-like order relation and $\{a_k\}_{k=1}^\infty$ be a sequence of positive numbers. For the existence of an F-widely spread sequence $\{x_k\}_{k=1}^\infty$ such that $\|x_k\| = a_k$, $k \in \mathbf{N}$, it is necessary and sufficient that $a_i^2 \leq \sum_{k \in F(i)} a_k^2$ for all $i \in \mathbf{N}$.*

PROOF. Necessity is a consequence of condition 1) of Definition 8.1.4. Sufficiency follows from the fact that we are working in an infinite-dimensional space and when we construct the elements $\{x_i\}_{i \in F^{n+1}(1)}$ we can add to $\mathrm{Lin}\{x_i\}_{i \in F^n(1)}$ enough dimensions to ensure that all requirements of definitions 8.1.2 and 8.1.4 are fulfilled. □

THEOREM 8.1.1. *Let H be an infinite-dimensional Hilbert space. Then for any sequence $\{a_k\}_{k=1}^{\infty}$ of positive numbers satisfying the conditions $\lim_{k\to\infty} a_k = 0$ and $\sum_{k=1}^{\infty} a_k^2 = \infty$ there exists a series $\sum_{k=1}^{\infty} z_k$ of elements of H such that $\|z_k\| = a_k$ for all $k \in \mathbf{N}$ and $\mathrm{WSR}(\sum_{k=1}^{\infty} z_k)$ is not a linear set. More precisely, for some rearrangement of terms the series converges to 0, for other it converges to some element $z \in H$, $z \neq 0$, and there is no rearrrangement for which the series converges weakly to tz with t not an integer.*

PROOF. Construct by induction a sequence of natural numbers $\{n_i\}$ as follows: $n_1 = 1$; n_{k+1} is the smallest number for which

$$\sum_{i=n_k+1}^{n_{k+1}} a_i^2 > 2a_k^2.$$

Define the tree-like order relation F by $F(k) = \{n_k + 1, n_k + 2, \dots, n_{k+1}\}$. Then F enjoys the following properties:

$$2a_i^2 \leq \sum_{k\in F(i)} a_k^2, \tag{3}$$

$$\sum_{i\in F^n(1)} a_i^2 \geq 2^n a_1^2, \tag{4}$$

$$\lim_{n\to\infty} \sum_{i\in F(n)} a_i^2 = 0. \tag{5}$$

Using Lemma 8.1.5, pick an F-widely spread sequence $\{x_i\}_{i=1}^{\infty} \subset H$ such that $\|x_i\| = a_i$, $i \in \mathbf{N}$. Now as the terms of the sought-for series $\sum_{i=1}^{\infty} z_i$ take $\pm x_i$ where the sign is "$-$" for $i \in F^{2n-1}(1)$, $n \in \mathbf{N}$, and "$+$" for the remaining $i \in \mathbf{N}$. If we write the elements z_i in the order

$$\sum_{n=0}^{\infty} \sum_{i\in F^{2n}(1)} \left(x_i - \sum_{k\in F(i)} x_k \right)$$

($F^0(1) = \{1\}$), where one first calculates the inner sums, and then the outer sum, then thanks to conditions 1) in definitions 8.1.2 and 8.1.4 and to relation (5), the series will converge in norm to 0. If now the series is rewritten in the order

$$x_1 + \sum_{n=1}^{\infty} \sum_{i\in F^{2n-1}(1)} \left(-x_i + \sum_{k\in F(i)} x_k \right),$$

then it will converge to x_1.

Let us show that if the series $\sum_{i=1}^{\infty} z_{\pi(i)}$ converges weakly to tx_1, then t is necessarily an integer. To this end let us denote by ε the distance from t to the nearest integer. Further, let S_n denote the partial sums of the series $\sum_{i=1}^{\infty} z_{\pi(i)}$, and put $D_n = S_n - tx_1$. By hypothesis, the sequence D_n converges weakly to 0. Since for m large enough we can write S_n as a linear combination $\sum_{i \in F^m(1)} t_i x_i$ with integer coefficients t_i, and since $x_1 = \sum_{i \in F^m(1)} x_i$, by assertion a) of Lemma 8.1.1 and Lemma 8.1.3 we have $\|T_k(D_n)\| \geq \varepsilon \|x_k\| = \varepsilon a_k$. Denote $\sup_n \|D_n\|$ by D. By Lemma 8.1.4,

$$D^2 \geq \liminf_{k \to \infty} \sum_{i \in F^n(1)} \|T_i(D_k)\|^2 \geq \varepsilon \sum_{i \in F^n(1)} a_i^2 \geq \varepsilon a_1^2 \cdot 2^n$$

for all $n \in \mathbf{N}$. Since n is arbitrary, this means that $\varepsilon = 0$ and t is an integer. □

REMARK 8.1.2. An examination of the preceding proof shows that no subsequence S_{n_k} of the sequence S_n of partial sums of the series $\sum_{i=1}^{\infty} z_{\pi(i)}$ can converge weakly to a point of the form tx_1 with t not an integer.

EXERCISE 8.1.1. Show that series with nonlinear weak sum range exists in any infinite-dimensional Banach space.

The authors cannot provide an example of a series in a Hilbert space with two-point weak sum range.

§2. Rearrangements of Series of Functions

In addition to the already familiar to the reader Problem 106 of the Scottish Book, solved by J. Marcinkiewicz, there is the following similar problem that goes back to S. Banach and W. Orlicz: does Steinitz's theorem extend to almost-everywhere convergence of series of functions? This turned to be a considerably more difficult problem; a counterexample was constructed by E. M. Nikishin [61] in 1971. In this section we will give a more convenient method for constructing similar examples, proposed by V. M. Kadets [39], which allows one to solve some other problems as well. Thus, we will construct a series of bounded functions on the segment $[0, 1]$ with the property that for one rearrangement of its terms the series converges uniformly to zero, for another rearrangement it converges uniformly to a function $u(t)$, but there is no rearrangement for which the series will converge, even in the sense of convergence in measure, to a third function, different from zero and $u(t)$.

Recall that for a subset $A \subset [0, 1]$ we denote by $\chi_A(t)$ the indicator function of A. The construction will make use of two classical orthonormal systems in $L_2[0, 1]$—the Rademacher system $r_n(t) = \operatorname{sign}(\sin(2^n \pi t))$, and the Haar system. The Rademacher system, which is the simplest example of a sequence of independent random variables that take the values ± 1 with equal probabilities, was already mentioned before. Let us recall the definition of the Haar system:

$$h_1(t) = \chi_{[0,1]}(t), \quad h_2(t) = \chi_{[0,\frac{1}{2}]}(t) - \chi_{[\frac{1}{2},1]}(t),$$

$$h_3(t) = \sqrt{2}\left(\chi_{(0,\frac{1}{4}]}(t) - \chi_{(\frac{1}{4},\frac{1}{2}]}(t)\right), \quad h_4(t) = \sqrt{2}\left(\chi_{(\frac{1}{2},\frac{3}{4}]}(t) - \chi_{(\frac{3}{4},1]}(t)\right)$$

$$\dots\dots\dots\dots\dots\dots\dots\dots\dots\dots\dots ,$$

$$h_{2^{n-1}+k}(t) = \sqrt{2^{n-1}}\left(\chi_{(\frac{2k-2}{2^n},\frac{2k-1}{2^n}]}(t) - \chi_{(\frac{2k-1}{2^n},\frac{2k}{2^n}]}(t)\right),$$

where $n \in \mathbf{N}$ and $1 \le k \le 2^{n-1}$. The Haar system is known to be a basis in $L_2[0,1]$.

Let us denote by LR the class of measurable functions on $[0,1]$ that are sums of series of the form $\sum_{n=1}^{\infty} c_n r_n(t)$, $c_k \in \mathbf{R}$, which converge in the metric of $L_2[0,1]$. Since the system $\{r_n\}_{n=1}^{\infty}$ is orthonormal, the convergence of such a series is equivalent to the condition $\sum_{n=1}^{\infty} c_n^2 < \infty$.

LEMMA 8.2.1. *Let I be an arbitrary measurable subset of the segment $[0,1]$ of measure $m(I) > 2/3$. Then there exists a constant $K = K(m(I)) \in \mathbf{R}^+$, such that the inequality*

$$\int_0^1 [f(x)]^2 dx \le K \sup_{x\in I}[f(x)]^2 \tag{1}$$

holds for all $f \in LR$.

PROOF. Consider the system of functions $r_{i,j}(x) = r_i(x)r_j(x)$, $i > j$, and adjoin to it the function $\mathbf{1}$ identically equal to 1. The resulting system is an orthonormal sequence on $[0,1]$. By the Parseval inequality,

$$m(I) = \int_0^1 [\chi_I(x)]^2 dx \ge \left[\int_0^1 \mathbf{1}\cdot\chi_I(x)dx\right]^2 + \sum_{i>j}\left[\int_0^1 r_{i,j}(x)\chi_I(x)dx\right]^2$$

$$= [m(I)]^2 + \sum_{i>j}\left[\int_I r_{i,j}(x)dx\right]^2.$$

Hence,

$$\sum_{i>j}\left[\int_I r_{i,j}(x)dx\right]^2 \le m(I) - [m(I)]^2. \tag{2}$$

Let $f(x) = \sum_{n=1}^{\infty} c_n r_n(x)$, where $c_n = 0$ for all n larger than or equal to some number N. Then using the relation $r_n \cdot r_n \equiv 1$ we have

$$\int_I [f(x)]^2 dx = m(I)\sum_{n=1}^{\infty} c_n^2 + 2\sum_{i>j} c_i c_j \int_I r_{i,j}(x)dx.$$

Estimating the last term by the Cauchy-Bunyakovskii inequality and applying inequality (2), we obtain

$$\int_I [f(x)]^2 dx \ge m(I)\sum_{n=1}^{\infty} c_n^2 - 2\left(\sum_{i>j}(c_ic_j)^2\right)^{1/2} \left(m(I) - [m(I)]^2\right)^{1/2} \ge$$

$$\geq m(I)\sum_{n=1}^{\infty} c_n^2 - \sqrt{2}\left(m(I) - [m(I)]^2\right)^{1/2}\sum_{n=1}^{\infty} c_n^2$$

$$= m(I)\left[1 - \left(2\,\frac{1-m(I)}{m(I)}\right)^{1/2}\right]\sum_{n=1}^{\infty} c_n^2. \tag{3}$$

By an obvious passage to the limit one can show that inequality (3) holds for arbitrary functions $f \in LR$, which are not necessarily finite linear combinations of Rademacher functions. Denote $1 - \left(2\,\frac{1-m(I)}{m(I)}\right)^{1/2}$ by K. Then

$$\sup_{x\in I}[f(x)]^2 \geq \frac{1}{m(I)}\int_I [f(x)]^2 dx \geq K\sum_{n=1}^{\infty} c_n^2 = K\int_0^1 [f(x)]^2 dx. \qquad \square$$

COROLLARY 8.2.1. *The class LR is closed in the sense of convergence in measure; moreover, any sequence of elements of LR that converges in measure also converges in the metric of $L_2[0,1]$.*

PROOF. Let $f_n(x)$ a sequence of functions from LR which converges in measure to a function $f(x)$. By the definition of convergence in measure, there exist measurable sets $I_n \subset [0,1]$, $m(I_n) \to 1$, for which

$$\lim_{n\to\infty}\sup_{x\in I_n}|f_n(x) - f(x)| = 0.$$

Denoting $I_n \cap I_m = I_{n,m}$ we obtain

$$\lim_{n,m\to\infty}\sup_{x\in I_{n,m}}|f_m(x) - f_n(x)| = 0 \quad \text{and} \quad \lim_{n,m\to\infty} m(I_{n,m}) = 1.$$

By Lemma 8.2.1, this means that $\{f_n\}$ is a Cauchy sequence in the metric of $L_2[0,1]$, and consequently it converges in $L_2[0,1]$ to some function $g(x)$. It is clear that $f(x) = g(x)$ almost everywhere and that f belongs to the closure of the class LR in $L_2[0,1]$. Hence, $f \in LR$. $\square$

LEMMA 8.2.2. *Let $f(x) = \sum_{n=1}^{\infty} c_n r_n(x)$. Then*

$$\sup_{x\in[0,1]}|f(x)| = \sum_{n=1}^{\infty}|c_n|;$$

moreover, if one of the sides of this relation is equal to infinity, then so is the second.

PROOF. It suffices to calculate directly the left-hand side of the equality. $\square$

THEOREM 8.2.1. *In the space l_1 there exists a series $\sum_{n=1}^{\infty} z_n$ such that for one rearrangement the series converges to zero, for another it converges to some $v \in l_1$, and there is no rearrangement for which the series converges, even in the metric of l_2, to a third element of l_2.*

PROOF. We shall make use of the construction of a series with two-point sum range given in §2 of Chapter 3; however, to avoid any confusion, the series constructed therein will be denoted here by $\sum_{n=1}^{\infty} d_n$.

Define a system of functions $c_{i,j}^n$ on the cube $Q = [0,1]^\omega$ as follows: $c_{1,1}^1 \equiv 1$, $c_{i,j}^n(t_1, t_2, \dots) = h_i(t_n)h_j(t_{n+1})$, where $i, j, n \in \mathbf{N}$, $i \neq 1$, and h_i are the Haar functions. These functions form an orthonormal system in $L_2(Q)$, and each term of the series $\sum_{n=1}^{\infty} d_n$ can be represented as a finite linear combination of functions $c_{i,j}^n$. Denote the closed linear span of the system $\{c_{i,j}^n\}$ by H. Define an isomorphism $T\colon H \to l_2$ by choosing some bijective correspondence between the unit basis vectors of l_2 and the functions $c_{i,j}^n$. Put $z_n = T(d_n)$, $v = T(c_{1,1}^1)$. Clearly, the series $\sum_{n=1}^{\infty} z_n$ has the property that for one rearrangement it converges in l_2 to 0, for another it converges to v, and there is no rearrangement for which it converges in l_2 to some element different from 0 and v.

The elements z_n are linear combinations of unit basis vectors, and so $z_n \in l_1$. We claim that for the rearrangements of the series $\sum_{n=1}^{\infty} d_n$ constructed in §2 of Chapter 3, for which that series converges to 0 or to 1, the series $\sum_{n=1}^{\infty} z_n$ converges in l_1 to 0 or to v, respectively. Indeed, for those rearrangements of the series $\sum_{n=1}^{\infty} d_n$, the difference between the sum and the partial sum has the form

$$R(t) = R(t_1, t_2, \dots) = \begin{cases} 1 & \text{if } (t_n, t_{n+1}) \in [a_1, b_1] \times [a_2, b_2] \\ 0 & \text{otherwise,} \end{cases}$$

where one of the segments $[a_1, b_1]$, $[a_2, b_2]$ has the form $[m/k, (m+1)/k]$, and the length $1/k$ of this segment tends to 0 when the index of the partial sum tends to infinity. Expanding $R(t)$ with respect to the functions $c_{i,j}^n$, we find that the sum of the absolute magnitudes of the coefficients of the expansion does not exceed $A \ln k/\sqrt{k}$. We thus conclude that the images under the map T of the rearrangements of the series $\sum_{n=1}^{\infty} d_n$ discussed in §2 of Chapter 3 converge not only in the metric of l_2, but also in the metric of l_1. □

THEOREM 8.2.2. *There exists a series $\sum_{n=1}^{\infty} f_n(x)$ of bounded measurable functions on $[0,1]$ such that for one rearrangement of its terms the series converges uniformly to zero, for another it converges uniformly to a function $u(x)$, but even with respect to convergence in measure there is no rearrangement for which it converges to a function that differs from zero and $u(x)$.*

PROOF. Let $P\colon l_2 \to LR$ denote the linear isomorphism defined by the rule $P((a_1, a_2, a_3, \dots)) = \sum_{n=1}^{\infty} a_n r_n$. Now take for the series $\sum_{n=1}^{\infty} f_n$ whose existence is asserted in the theorem the image under P of the series $\sum_{n=1}^{\infty} z_n$ constructed in Theorem 8.2.1: $f_n = P(z_n)$. By Lemma 8.2.2, the functions $f_n(x)$ are bounded on

$[0,1]$, and if for some permutation π the series $\sum_{n=1}^{\infty} z_{\pi(n)}$ converges in l_1, then the series $\sum_{n=1}^{\infty} f_{\pi(n)}(x)$ converges uniformly on $[0,1]$. Hence, by Theorem 8.2.1, there exist two rearrangements of the series $\sum_{n=1}^{\infty} f_n$ for which it converges uniformly to zero and to $u = P(v)$, respectively. To complete the proof it remains to use Lemma 8.2.1 and the orthonormality of the sequence $\{r_n(x)\}$. □

REMARK 8.2.1. If instead of the Rademacher functions we use the lacunary trigonometric system $\{\sin(2^n \pi x)\}$, we can take the functions $f_n(x)$ in Theorem 8.2.2 to be trigonometric polynomials.

EXERCISE 8.2.1. In his PhD dissertation P. A. Kornilov has shown that if one expands the function in Marcinkiewicz's example with respect to the orthonormal Haar systems, then that example works in the space l_1. What results on the sum range of series of functions can be derived from this?

EXERCISE 8.2.2. From Corollary 8.2.1 one can obtain another proof of the Khinchin inequalities for $1 \leq p \leq 2$ (and even for $0 < p < 1$!). One can also prove Lemma 8.2.1, Corollary 8.2.1, and Khinchin inequalities for the functions $\sin(2^n \pi x)$ instead of the Rademacher functions. Do it!

§3. Banaszczyk's Theorem on Series in Metrizable Nuclear Spaces

For this section we are assuming that the reader is familiar with the basic definitions of the theory of locally convex spaces (if not, she is referred to the books [45, Chapter 3, §5], [76] or [78], which will suffice for our purposes). On the other hand, the notion of nuclear space, despite its extreme usefulness (for instance, the spaces of analytic functions with the topology of convergence on compact sets, the spaces of infinitely differentiable functions, and others, are nuclear), is still insufficiently popular, and is usually not incorporated in the university courses. A good (even if somewhat old) source for learning the theory of nuclear spaces is A. Pietsch's monograph [69]. To facilitate the understanding of the ensuing material we shall give all the necessary definitions.

Let E be a metrizable locally convex linear space, i.e., the topology of E is given by a countable family of seminorms $p_1 \leq p_2 \leq p_3 \leq \dots$. Denote by $B_n = B_{p_n}$ the unit balls with respect to these seminorms. Convergence in E means convergence in all the seminorms p_n. A continuous linear functional $f \in E'$ is a linear functional that is continuous with respect at least one of the seminorms p_n, or, equivalently, bounded on at least one of the balls B_n. For series of elements of E we preserve the terminology introduced earlier for the Banach space case: SR denotes the sum range, $\Gamma \subset E'$—the set of convergence functionals, and $\Gamma^{\perp} \subset E$—the annihilator of Γ.

We cannot indicate a specific author for the following theorem, since apparently every mathematician who has given some thought to the possibility of generalizing Steinitz's theorem to the case of topological vector spaces did discover this result on his own. With the hope that the reader will be able to provide independently the necessary arguments, we will give here only the statement of the theorem, omitting the proof.

THEOREM 8.3.1. *Suppose that in the metrizable locally convex space E the following conditions are satisfied:*

(A) Rounding-off-Coefficients Lemma. *For any continuous seminorm p on E there exists a continuous seminorm $q \geq p$ with the property that, for any elements $x_1, \ldots, x_n$, $n \in \mathbf{N}$, of the ball B_q and any set of coefficients $\mu_1, \ldots, \mu_n$, $0 \leq \mu_i \leq 1$, one can find coefficients $\theta_i \in \{0, 1\}$ such that*

$$p\left(\sum_{i=1}^{n} \mu_i x_i - \sum_{i=1}^{n} \theta_i x_i\right) \leq 1.$$

(B) Permutation Lemma. *For any continuous seminorm p on E there exists a continuous seminorm $q \geq p$ with the property that, for any finite set of elements $x_1, \ldots, x_n$ of the ball B_q satisfying $\sum_{i=1}^{n} x_i = 0$, one can choose a permutation π of the first n natural numbers such that*

$$\max_{k \leq n} p\left(\sum_{i=1}^{k} x_{\pi(i)}\right) \leq 1.$$

Then for any conditionally convergent series $\sum_{k=1}^{\infty} x_k$ in the space E one has that $\mathrm{SR}(\sum_{k=1}^{\infty} x_k) = a + \Gamma^{\perp}$, where $a = \sum_{k=1}^{\infty} x_k$. □

EXERCISE 8.3.1. Show that in the above theorem condition (A) is a consequence of condition (B), and hence the statement of the theorem can be simplified.

DEFINITION 8.3.1. Let A and B be two convex subsets of a linear space E. The *Kolmogorov widths* of the set A with respect to the set B are the numbers

$$d_k(A, B) = \inf_{L, x} \inf\{\varepsilon > 0 \colon A \subset x + L + \varepsilon B\},$$

where the second infimum is taken over all subspaces $L \subset E$ with $\dim L < k$ and over all $x \in E$.

Notice that $d_k(A, B)$ does not change when the sets A and B undergo parallel translation, does not decrease when A is enlarged, and does not increase when B is enlarged; moreover, d_k does not increase when k is increased.

In what follow U_n will denote the standard Euclidean ball in $\mathbf{R}^n$:

$$U_n = \{x = (x_1, \ldots, x_n) \colon \sum_{i=1}^{n} |x_i|^2 \leq 1\}.$$

EXERCISE 8.3.2. Show that if A is an ellipsoid in $\mathbf{R}^n$ with principal semi-axes $\lambda_1 \geq \lambda_2 \geq \ldots \geq \lambda_n$, then $d_k(A, U_n) = \lambda_k$.

In what follows orthogonality in $\mathbf{R}^n$ will be understood as orthogonality with respect to the standard inner product, associated with the ball U_n.

DEFINITION 8.3.2. A seminorm p on a linear space E is said to be *Euclidean* if it is given by some nonnegative bilinear form $Q(x, y)$: $p(x) = (Q(x, x))^{1/2}$.

In a finite-dimensional space the unit ball of such a seminorm is either an ellipsoid, or an elliptic cylinder, i.e., the direct sum of an ellipsoid and a subspace.

DEFINITION 8.3.3. A space E is said to be *countably Euclidean* if its topology can be given by a countable family of Euclidean norms.

DEFINITION 8.3.4. A countably Euclidean space E is said to be *nuclear* if for any continuous Euclidean seminorm p on E there exists a continuous Euclidean seminorm q such that

$$\sum_{k=1}^{\infty} d_k^2(B_q, B_p) \leq 1.$$

Since we are not dealing here with the nonmetrizable case, in what follows countably-Euclidean nuclear spaces will be simply referred to as nuclear spaces.

REMARK 8.3.1. In Definition 4.3.3 we introduced the concept of a nuclear space in a somehwat different way. In the metrizable case the two definitions are equivalent; in particular, any nuclear space in the sense of Definition 4.3.3 is countably Euclidean. For more on this see [69].

The main goal of this section is to show that Steinitz's theorem holds in nuclear spaces. We shall follow the original proof of this result, obtained by W. Banaszczyk in 1989 [2].

LEMMA 8.3.1. *Let C be an n-dimensional ellipsoid in $\mathbf{R}^n$ with principal semi-axes $\lambda_1, \ldots, \lambda_n$, and let P be an n-dimensional right-angled parallelepiped circumscribed to C. Then*

$$\operatorname{diam} P = 2\left(\sum_{k=1}^{n} \lambda_k^2\right)^{1/2}.$$

PROOF. We may assume that the center of C is at the origin. Let $T\colon \mathbf{R}^n \to \mathbf{R}^n$ be a linear operator for which $T(U_n) = C$. Pick an orthonormal basis $\{e_k\}_{k=1}^n$ in $\mathbf{R}^n$ whose vectors are parallel to the edges of the parallelepiped P. Then

$$P = \left\{\sum_{k=1}^{n} t_k e_k \colon |t_k| \leq s_k,\ k = 1, \ldots, n\right\},$$

where $s_k > 0$ is half of the length of the corresponding edge. By hypothesis,

$$s_k = \sup_{u \in U_n} \langle Tu, e_k \rangle = \sup_{u \in U_n} \langle u, T^* e_k \rangle = \|T^* e_k\|$$

and

$$\frac{1}{2} \operatorname{diam} P = \left(\sum_{k=1}^{n} s_k^2 \right)^{1/2} = \left(\sum_{k=1}^{n} \|T^* e_k\|^2 \right)^{1/2}.$$

It remains to observe that the right-hand side in the last equality is precisely the Hilbert-Schmidt norm of the operator T, and hence is equal to $\left(\sum_{k=1}^{n} \lambda_k^2\right)^{1/2}$. □

LEMMA 8.3.2. *Let $C \subset \mathbf{R}^n$ be an n-dimensional ellipsoid centered at the origin whose principal semi-axes $\lambda_1, \dots, \lambda_n$ satisfy the condition*

$$\lambda_1^{-2} + \cdots + \lambda_n^{-2} \le 1, \tag{1}$$

and let M be an $(n-1)$-dimensional affine hyperplane that intersects U_n. Then $C \cap M$ is an $(n-1)$-dimensional ellipsoid whose principal semi-axes $\mu_1, \dots, \mu_{n-1}$ satisfy the condition

$$\mu_1^{-2} + \cdots + \mu_{n-1}^{-2} \le 1. \tag{2}$$

PROOF. By condition (1), $U_n \subset C$, and consequently $C \cap M \neq \emptyset$. Label the principal semi-axes of C in increasing order: $\lambda_1 \le \cdots \le \lambda_n$. Then

$$\lambda_k^{-1} = d_k(U_n, C). \tag{3}$$

Similarly, denoting by u the center of the ellipsoid $C \cap M$, setting $M_0 = M - u$, and labeling the semi-axes $\mu_1, \dots, \mu_{n-1}$ in increasing order, we obtain

$$\mu_k^{-1} = d_k(U_n \cap M_0, C \cap M). \tag{4}$$

Let $T\colon \mathbf{R}^n \to \mathbf{R}^n$ be a linear operator such that $T(C) = U_n$. Denote $N = T(M)$, $N_0 = T(M_0)$, $D = T(U_n)$, and $\nu_k = \lambda_k^{-1}$; then $\nu_1, \dots, \nu_n$ are the principal semi-axes of the ellipsoid D. By condition (1),

$$\nu_1^2 + \cdots + \nu_n^2 \le 1. \tag{5}$$

Let P be a right-angled parallelepiped circumscribed to D such that one of the $(n-1)$-dimensional faces of P is parallel to N. From (5) and Lemma 8.3.1 it follows that $P \subset U_n$. By construction, $D \subset P$ and $N \cap D \neq \emptyset$. Let Π denote the orthogonal projection of $\mathbf{R}^n$ onto the hyperplane N. Then $\Pi(P)$ is an $(n-1)$-dimensional right-angled parallelepiped circumscribed to the ellipsoid $\Pi(D)$; moreover, $\Pi(P) \subset$

$U_n \cap N$. If we equip N with a new Euclidean metric in which $U_n \cap N$ is the unit ball, then in this metric $\operatorname{diam} \Pi(P) \leq 2$, and, by Lemma 8.3.1,

$$\sum_{k=1}^{n-1} d_k^2(\Pi(D), U_n \cap N) \leq 1, \tag{6}$$

because $d_k(\Pi(D), U_n \cap N)$ is the length of the k-th principal semi-axis of the ellipsoid $\Pi(D)$ in the new metric.

Since $\Pi(D \cap N_0)$ can be obtained from $D \cap N_0$ by parallel translation and since $\Pi(D \cap N_0) \subset \Pi(D)$, inequality (6) gives

$$\sum_{k=1}^{n-1} d_k^2(D \cap N_0, U_n \cap N) \leq 1. \tag{7}$$

Next, since T is an isomorphism, $d_k(TA, TB) = d_k(A, B)$ for any two sets A, B. Therefore,

$$d_k(D \cap N_0, U_n \cap N) = d_k(U_n \cap M_0, C \cap M),$$

which in view of relations (4) and (7) yields the needed inequality (2). □

LEMMA 8.3.3. *Let P be an n-dimensional parallelepiped in $\mathbf{R}^n$ such that the lenghts of its edges are no larger than 1, let C be an n-dimensional ellipsoid in the same space whose principal semi-axes $\lambda_1, \ldots, \lambda_n$ obey the inequality $\lambda_1^{-2} + \cdots + \lambda_n^{-2} \leq 1$, and suppose the center of C lies in P. Then one of the edges of P is contained in C.*

PROOF. With no loss of generality we may assume that C is centered at the origin. Then $0 \in P$ and $0 \in C$. To prove the lemma we proceed by induction on the dimension n of the space. For $n = 1$ there is nothing to prove. Suppose the assertion of the lemma is already proved in dimension $n - 1$ and let us show that it holds true in dimension n. Denote $\sup\{r > 0 \colon rC \subset P\}$ by s. Then the ellipsoid sC is tangent to one of the $(n-1)$-dimensional faces F of the parallelepiped P at some point w. Let M be the hyperplane containing F. Since $0 \in P$ and the lengths of the edges of P are not larger than 1, it follows that $U_n \cap M \neq \emptyset$. By Lemma 8.3.2, the pricipal semi-axes $\mu_1, \ldots, \mu_{n-1}$ of the ellipsoid $C \cap M$ satisfy the condition $\mu_1^{-2} + \cdots + \mu_{n-1}^{-2} \leq 1$. Since w is the center of the ellipsoid $C \cap M$ and $w \in F$, the inductive hypothesis says that one of the edges of the parallelepiped F is contained in $C \cap M$. But this edge is also an edge of the original parallelepipied P and lies in C. □

Let us denote by I_s the collection of all subsets of the set $\{1, \ldots, s\}$.

LEMMA 8.3.4 (ROUNDING-OFF-COEFFICIENTS LEMMA IN NUCLEAR SPACES). *Let $p \geq q$ be two Euclidean seminorms on a vector space E satisfying*

$$\sum_{k=1}^{\infty} d_k^2(B_p, B_q) \leq 1. \tag{8}$$

Let $b \in E$, $v_1, \dots, v_s \in B_p$, $s \geq 2$, and let $A = \{b + \sum_{i=1}^{s} t_i v_i : 0 \leq t_i \leq 1\}$. Suppose $0 \in A$. Then there exists $J \in I_s$, $1 \leq \operatorname{card} J \leq s-1$, such that $b + \sum_{i \in J} v_i \in B_q$.

PROOF. Since we are dealing with finitely many vectors, confining our considerations to their linear span we may assume that the space E is finite-dimensional. Further, with no loss of generality we may assume that B_p and B_q are ellipsoids (and not elliptic cylinders) and that the vectors $\{v_i\}_{i=1}^{s}$ are linearly independent (the general case is readily derived from this case by a limit process). Finally, choosing an appropriate isomorphism we may assume that $E = \mathbf{R}^s$ and $B_p = U_s$. Then condition (8) says that the principal semi-axes $\lambda_1, \dots, \lambda_n$ of the ellipsoid B_q satisfy the inequality $\lambda_1^{-2} + \dots + \lambda_s^{-2} \leq 1$. The set A is a parallelepiped that contains the point 0, i.e., the center of the ellipsoid B_q. By Lemma 8.3.3, one of the edges of A is contained in B_q. Since $s \geq 2$, one of the vertices of that edge is different from both b and $b + \sum_{i=1}^{s} v_i$. Since every vertex of the parallelepiped A has the form $b + \sum_{i \in J} v_i$, this completes the proof of the lemma. □

REMARK 8.3.2. Here the Rounding-off-Coefficients Lemma is just an intermediate step in the proof of the Permutation Lemma in a nuclear space. Otherwise, in view of Exercise 8.3.1 we would not need this lemma.

LEMMA 8.3.5. *Let $B \subset C \subset D$ be three n-dimensional ellipsoids with common center (the origin) and such that*

$$\sum_{k=1}^{n} d_k^2(B, C) \leq 1 \tag{9}$$

and

$$\sum_{k=1}^{n} d_k^2(C, D) \leq \frac{1}{4}. \tag{10}$$

Then, given arbitrary elements $u_1, \dots, u_s \in B$ and $a \in C$ satisfying $a + \sum_{i=1}^{s} u_i \in C$, one can choose a permuation σ of the set $\{1, \dots, s\}$ such that

$$a + \sum_{i=1}^{j} u_{\sigma(i)} \in D, \quad j = 1, \dots, s. \tag{11}$$

PROOF. We proceed by simultaneous induction on n and s. For the cases ($n = 1$, $s \in \mathbf{N}$ arbitrary) and ($s = 1$, $n \in \mathbf{N}$ arbitrary) the assertion of the lemma is obvious. Suppose it is true for all pairs $(\overline{n}, \overline{s})$ such that $\overline{n} < n$ and $\overline{s} \leq s$, or $\overline{n} = n$ and $\overline{s} < s$, and let us verify that it is true for the pair (n, s). Thus, we need to establish the existence of a permutation σ such that (11) holds. We shall consider two separate cases:

1) Suppose $0 \in A = \{a + \sum_{i=1}^{s} t_i u_i \colon 0 \le t_i \le 1\}$. Then, by Lemma 8.3.4, there exists a set $J \in I_s$, $1 \le \operatorname{card} J \le s-1$, for which

$$a + \sum_{i \in J} u_i \in C. \tag{12}$$

For the sake of convenience we shall assume, with no loss of generality, that $J = \{1, \dots, z\}$, with $z \le s-1$. Applying the inductive hypothesis to the set J we obtain a permutation σ of J such that

$$a + \sum_{i=1}^{j} u_{\sigma(i)} \in D, \quad j = 1, \dots, z.$$

Hence, it remains to permute in the needed way the elements $u_{z+1}, \dots, u_s$. If we denote $a + \sum_{i=1}^{z} u_i$ by b, then $b \in C$ (as shown by relation (12)), and $b + \sum_{i=z+1}^{s} u_i \in C$ (by hypothesis). Applying again the inductive hypothesis, but now to the indices $\{z+1, \dots, s\}$, we extend the permutation σ to this set of indices in such a manner that

$$b + \sum_{i=z+1}^{j} u_{\sigma(i)} = a + \sum_{i=1}^{j} u_{\sigma(i)} \in D, \quad j = z+1, \dots, s.$$

This completes the construction of the required permutation σ.

2) Suppose now that $0 \notin A$. Then there exists a linear functional f on $\mathbf{R}^n$ such that

$$f(u) > 0 \text{ for all } u \in A. \tag{13}$$

Denote $h = \sup\{f(u) \colon u \in C\}$. The set A is centrally symmetric about the point $x = a + \frac{1}{2}\sum_{i=1}^{s} u_i$. Since x is the midpoint of the segment connecting a and $a + \sum_{i=1}^{s} u_i$, it follows that $x \in C$. Next, since the center of symmetry of A lies in C, condition (13) implies that

$$\sup\{f(u) \colon u \in A\} \le 2h. \tag{14}$$

With no loss of generality we may assume that $D = U_n$. Let $H = \operatorname{Ker} f$ (then $\dim H = n-1$), and let Π denote the orthogonal projection onto H. Inequality (9) obviously yields

$$\sum_{k=1}^{n-1} d_k^2(\Pi(B), \Pi(C)) \le 1. \tag{15}$$

Circumscribe an n-dimensional right-angled parallelepiped P to C such that one of the $(n-1)$-dimensional faces of P is parallel to H. By Lemma 8.3.1 and relation (10), $2P \subset U_n$. From this and also from the construction of P is follows that

$$2\Pi(P) \subset W \overset{\text{def}}{=} \Pi(U_n \cap f^{-1}(2h)). \tag{16}$$

$\Pi(P)$ is an $(n-1)$-dimensional parallelepiped circumscribed to the ellipsoid $\Pi(C)$. By Lemma 8.3.1, inclusion (16) means that

$$\sum_{k=1}^{n-1} d_k^2(\Pi(C), W) \leq \frac{1}{4}. \tag{17}$$

Let us apply the inductive hypothesis to the ellipsoids $\Pi(B) \subset \Pi(C) \subset W$ and the elements b and $\{v_i\}_{i=1}^n$, defined as $b = \Pi(a)$, $v_i = \Pi(u_i)$ (that we are allowed to do this is ensured by inequalities (15) and (17)). We obtain a permutation σ such that

$$\Pi(a) + \sum_{i=1}^{j} \Pi(u_{\sigma(i)}) \in W, \quad j = 1, \ldots, s.$$

By relations (13) and (14), this means that

$$a + \sum_{i=1}^{j} u_{\sigma(i)} \in \Pi^{-1}(W) \cap f^{-1}([0, 2h]),$$

which by the definition of the set W implies that (11) holds. □

Using the fact that the Permutation Lemma deals with a finite number of vectors, we can derive from the preceding finite-dimensional variant the needed Permutation Lemma in nuclear spaces, which in turn implies *Banaszczyk's theorem* (see [2], [3]):

THEOREM 8.3.2. *In metrizable nuclear spaces the assertion of Steinitz's theorem on the sum range of conditionally convergent series holds true.*

EXERCISE 8.3.3. Show that in the nonmetrizable nuclear space $\mathbf{R}^{[0,1]}$ Steinitz's theorem fails.

EXERCISE 8.3.3. Give an example of a nonmetrizable nuclear space in which Steinitz's theorem does hold.

APPENDIX
THE LIMIT SET OF THE RIEMANN INTEGRAL SUMS OF A VECTOR-VALUED FUNCTION

Consider a bounded function $f\colon [0,1] \to X$, where X is a Banach space. Denote $M(f) = \sup\{\|f(t)\|, t \in [0,1]\}$. Let $\Gamma = \{\Delta_i\}_{i=1}^n$ be a partition of the segment $[0,1]$ into segments $\Delta_i = [a_i, b_i]$, $0 = a_1 < b_1 = a_2 < b_2 = a_3 < b_3 = \ldots < b_n = 1$. Take a set of points $T = \{t_i\}_{i=1}^n$, $t_i \in \Delta_i$. It is natural to define the Riemann integral sum of the function f as

$$S(f, \Gamma, T) = \sum_{i=1}^{n} f(t_i)(b_i - a_i) = \sum_{i=1}^{n} f(t_i)|\Delta_i|,$$

where $|\Delta_i|$ denotes the length of the segment Δ_i.* Call $d(\Gamma) = \max_i |\Delta_i|$ the *diameter* of the partition Γ. As in the usual real-valued case, we will say that f is *Riemann integrable* if the integral sums $S(f, \Gamma, T)$ tend to some limit when $d(\Gamma) \to 0$. The limit is called the *integral* of the function f.

If for real-valued functions Riemann integrability is a very restrictive condition and the Lebesgue integral, thanks to its considerably more general nature, turns out to be an incomparably more convenient object, for Banach space-valued functions the situation is more complicated.

EXAMPLE 1**. Let $l_2[0,1]$ be the space of real-valued functions that are different from zero in at most countably many points of the segment $[0,1]$ and obey the condition $\sum_{t\in[0,1]} |g(t)|^2 < \infty$. Equip this space with the norm $\|g\| = \left(\sum_{t\in[0,1]} |g(t)|^2\right)^{1/2}$. Defined in this way, $l_2[0,1]$ is a nonseparable Hilbert space with a noncountable orthonormal basis e_x, $x \in [0,1]$, where $e_x(t) = 1$ if $t = x$, and $e_x(t) = 0$ if $t \neq x$. Consider the function $f\colon [0,1] \to l_2[0,1]$ defined by $f(x) = e_x$. Then any integral sum of f is estimated as

$$\|S(f, \Gamma, T)\| = \left\| \sum_{i=1}^{n} e_{t_i} |\Delta_i| \right\| = \left(\sum_{i=1}^{n} |\Delta_i|^2 \right)^{1/2}$$

---

*To keep in agreement with the notation $\int f\,dx$ for the integral, we will allow here vectors to be multiplied by scalars to the right.

**The numbering in the Appendix is independent from the one in the main text.

$$\leq \left(\max_i |\Delta_i|\right)^{1/2} \left(\sum_{i=1}^{n} |\Delta_i|\right)^{1/2} = (d(\Gamma))^{1/2}.$$

We see that when $d(\Gamma) \to 0$ the integral sum also tends to zero, and hence the constructed function f is Riemann integrable. At the same time f cannot be represented as the limit of an almost everywhere convergent sequence of simple functions (i.e., functions, each of which takes only a finite number of values). Thus, we produced an example of a function that is Riemann integrable, but not measurable. As a matter of fact, our Riemann-integrable function is discontinous at any point.

For functions that are not Riemann integrable the role of an integral may be played by the limit of some sequence $S(f, \Gamma_n, T_n)$ of integral sums, where the partitions Γ_n are increasingly finer: $d(\Gamma_n) \to 0$ as $n \to \infty$. For a nonintegrable function there may be many such limits. It therefore becomes necessary to consider not a single limit, but rather the set of all such limits.

DEFINITION 1. For a bounded function $f\colon [0,1] \to X$ we call the set of all points of X that can be represented as the limit of a sequence of integral sums $S(f, \Gamma_n, T_n)$ with $d(\Gamma_n) \to \infty$ the *set of limits of Riemann integral sums of the function* f and denote it by $\mathrm{I}(f)$.

Obviously, $\mathrm{I}(f)$ is a closed set contained in the ball of radius $M(f)$.

In this appendix we will discuss various results connected with the description of the structure of the set $\mathrm{I}(f)$. This topic has many aspects that are similar to results and methods of the theory of conditionally convergent series, but is considerably less developed and does not have such a definitive character as the latter.

§1. Functions Valued in a B-Convex Space

For real-valued functions f the set $\mathrm{I}(f)$ is the segment whose endpoints are the lower and upper Riemann integrals. This simple fact is readily generalized to the case of functions taking values in a finite-dimensional space: compactness considerations show that $\mathrm{I}(f)$ is a nonempty connected set. The following theorem, proved in 1947 by P. Hartman [3],* should be regarded as the first nontrivial result from which the theory expounded below originated: for any bounded function taking values in a finite-dimensional space the set $\mathrm{I}(f)$ is convex. This theorem plays the same role in the theory under consideration as Steinitz's theorem in the theory of series. However, in contrast to Steinitz's theorem, Hartman's theorem admits a direct generalization to some infinite-dimensional spaces. Thus, in 1954 I. Halperin and N. Miller [2] extended Hartman's theorem to functions with values in a Hilbert space. Developing these ideas, in 1966 M. Nakamura and I. Amemiya [10] generalized the result to B-convex spaces. Since the proofs in all these cases

*In the Appendix we will always refer to the list of "Supplementary References."

can be carried out by completely similar means, here we shall prove directly the last, most general version, following in our exposition the paper [8], where the Nakamura-Amemiya theorem has been rediscovered (with a delay of 20 years!) by the authors of the present book.

THEOREM 1. *Let X be a B-convex normed space and $f\colon [0,1] \to X$ be a bounded function. Then the set* $\mathrm{I}(f)$ *is convex.*

PROOF. Recall (Chapter 5, §3) that X, being a B-convex space, has some infratype $p > 1$, i.e., there exists a constant $C > 0$ such that the inequality

$$\min_{\alpha_i=\pm 1}\left\|\sum_{i=1}^{n}\alpha_i x_i\right\| \le C\left(\sum_{i=1}^{n}\|x_i\|^p\right)^{1/p} \tag{1}$$

holds for any finite collection $\{x_i\}_{i=1}^{n}$ of elements of X.

Let x_1, x_2 be two points in $\mathrm{I}(f)$. Since the set $\mathrm{I}(f)$ is closed, to prove the theorem it suffices to show that $\frac{1}{2}(x_1+x_2) \in \mathrm{I}(f)$. To this end fix $N \in \mathbf{N}$ and choose sequences of partitions and of collections of points (Γ_n^1, T_n^1) and (Γ_n^2, T_n^2) such that

$$\lim_{n\to\infty} d(\Gamma_n^i) = 0, \quad S(f,\Gamma_n^i,T_n^i) \to x_i, \quad i = 1,2.$$

With no loss of generality one can assume that both partitions Γ_n^1 and Γ_n^2 include as division points all points of the form k/N, $k = 0,1,\dots,N$ (indeed, by adjoining these points to Γ_n^i one adds N segments whose lenghts do not exceed $d(\Gamma_n^i)$, and hence, for any way of selecting the necessary supplementary points T_n^i, the integral sum is modified by a term bounded by $2Nd(\Gamma_n^i)M(f)$, which tends to zero when $n \to \infty$). Pick $\varepsilon > 0$ and denote by S^1 one of the integral sums $S(f,\Gamma_n^1,T_n^1)$ for which $\|S^1 - x_1\| < \varepsilon$, and similarly take S^2 such that $\|S^2 - x_2\| < \varepsilon$. Next, divide the segment $[0,1]$ into N segments $\{\Delta_k\}_{k=1}^{N}$ of equal length and denote by S_k^i, $i = 1,2$, the part of the integral sum S^i corresponding to the segments of the partition that lie in Δ_k. Then $S^i = \sum_{k=1}^{N} S_k^i$, $i = 1,2$. If now for each of the segments Δ_k we choose in arbitrary manner either the sum S_k^1 or the sum S_k^2, we obtain 2^N different integral sums of the function f. These sums can be formally written as

$$S\left(\{\alpha_k\}_{k=1}^{N}\right) = \sum_{k=1}^{N}\left(\frac{1+\alpha_k}{2}S_k^1 + \frac{1-\alpha_k}{2}S_k^2\right),$$

where $\alpha_k = \pm 1$ are arbitrary. We claim that among these sums there is one that lies close enough to $\frac{1}{2}(x_1+x_2)$. Indeed,

$$\left\|\frac{1}{2}(x_1+x_2) - S\left(\{\alpha_k\}_{k=1}^{N}\right)\right\| \le \varepsilon + \left\|\frac{1}{2}(S^1+S^2) - S\left(\{\alpha_k\}_{k=1}^{N}\right)\right\|$$

$$= \varepsilon + \left\|\sum_{k=1}^{N}\frac{1}{2}(S_k^1+S_k^2) - \sum_{k=1}^{N}\left(\frac{1+\alpha_k}{2}S_k^1 + \frac{1-\alpha_k}{2}S_k^2\right)\right\|$$

$$= \varepsilon + \frac{1}{2} \left\| \sum_{k=1}^{N} \alpha_k (S_k^2 - S_k^1) \right\| .$$

Using inequality (1) and the relation $\|S_k^i\| \le M(f)/N$, we obtain

$$\min_{\alpha_k = \pm 1} \left\| \frac{1}{2}(x_1 + x_2) - S\left(\{\alpha_k\}_{k=1}^N\right) \right\| \le \varepsilon + CN^{\frac{1}{p}-1} M(f).$$

Since ε can be made arbitrarily small and N arbitrarily large, the last relation means precisely that the point $\frac{1}{2}(x_1 + x_2)$ can be arbitrarily well approximated by arbitrarily fine integral sums of the function f. □

We wish to emphasize an important feature of the above theorem: we have shown that the set $\mathrm{I}(f)$ is convex, but nothing was said about whether this set is nonempty. This question will be addressed in Section 3 of the Appendix.

§2. The Example of Nakamura and Amemiya

In functional analysis one often encounters theorems that generalize some or another result of classical analysis to functions with values in a Banach space. For some theorems of analysis such a generalization turns out to be possible for functions taking values in any space, regardless of the structure of that space. In other cases, however, whether the generalization is possible depends on the geometry of the space. It is precisely this situation that is of interest in Banach space theory, since any such theorem of analysis becomes a tool for investigating the structural properties of Banach spaces. From this point of view, the example constructed in the aforementioned paper [10] of a space for which Theorem 1 fails is of exceptional importance. This example, which will be given here is a somewhat modified form, yields the "limiting case" of nonconvex set: for the function f in question, $\mathrm{I}(f)$ consists of exactly two points.

Recall that $l_1[0,1]$ is the space of real-valued functions defined on the segment $[0,1]$ and such that $\sum_{\gamma \in [0,1]} |g(\gamma)| < \infty$ (from this condition it follows, in particular, that the function g takes nonzero values for at most countably many points). The norm in $l_1[0,1]$ is given by the formula $\|g\| = \sum_{\gamma \in [0,1]} |g(\gamma)|$. The unit basis vectors of the space $l_1[0,1]$ are defined to be the elements e_t given by

$$e_t(\gamma) = \begin{cases} 1 & \text{if } \gamma = t \\ 0 & \text{if } \gamma \ne t. \end{cases}$$

Then $\|e_t\| = 1$ for all $t \in [0,1]$. Any element $g \in l_1[0,1]$ can be written in the form $g = \sum_{k=1}^{\infty} a_k e_{t_k}$, and $\| \sum_{k=1}^{\infty} a_k e_{t_k} \| = \sum_{k=1}^{\infty} |a_k|$.

EXAMPLE 2. For any $n \in \mathbf{N}$ consider the finite subsets E_n and G_n of the segment $[0,1]$ defined as follows: E_n consists of the 2^n points of the form

$$\frac{k}{2^{n-1}} \pm \frac{\sqrt{2}}{8^n}, \quad k = 0, 1, 2, \ldots, 2^{n-1},$$

while G_n consists of the 2^n points of the form

$$\frac{k}{2^{n-1}} \pm \frac{\sqrt{2}}{10^n}, \quad k = 0, 1, 2, \ldots, 2^{n-1}$$

(we have 2^n and not $2^n + 1$ points, because two of the points of the indicated form fall outside $[0, 1]$).

Define the function $f\colon l_1[0,1] \to l_1[0,1]$ by the rule

$$f(t) = \begin{cases} e_t & \text{if } t \notin (\bigcup_n E_n) \cup (\bigcup_n G_n) \\ e_0 + \frac{1}{2^n} \sum_{\alpha \in E_n} e_\alpha - e_t & \text{if } t \in E_n \\ e_1 + \frac{1}{2^n} \sum_{\alpha \in G_n} e_\alpha - e_t & \text{if } t \in G_n. \end{cases}$$

Clearly, f is bounded and $M(f) \leq 3$. If we partition the segment $[0, 1]$ into 2^n equal segments (call this partition Γ_n), then each of them contains exactly one point from E_n. It follows that

$$S(f, \Gamma_n, E_n) = \sum_{t \in E_n} \frac{1}{2^n} f(t) = \frac{1}{2^n} \sum_{t \in E_n} \left(e_0 + \frac{1}{2^n} \sum_{\alpha \in E_n} e_\alpha - e_t \right) = e_0.$$

Similarly, $S(f, \Gamma_n, G_n) = e_1$. Hence, the points e_0 and e_1 belong to $\mathrm{I}(f)$.

THEOREM 2. *For the function f constructed in Example 2 the set $\mathrm{I}(f)$ consists exactly of the two points e_0 and e_1.*

PROOF. Since any unit vector e_α with $\alpha \neq 0, 1$ appears in the expression of only finitely many values of f, the expansion of an arbitrary $x \in \mathrm{I}(f)$ may contain only the vectors e_0 and e_1. In other words, any element $x \in \mathrm{I}(f)$ has the form $\lambda e_0 + \mu e_1$. We need to show that either $\lambda = 1$ and $\mu = 0$, or $\lambda = 0$ and $\mu = 1$.

Suppose that some sequence $S(f, \Gamma'_n, T'_n)$ of integral sums of the function f, with $\lim_{n\to\infty} d(\Gamma'_n) = 0$, converges to $\lambda e_0 + \mu e_1$. Passing to a subsequence (Γ_n, T_n) of the sequence (Γ'_n, T'_n), one can ensure that the integral sums $S_n = S(f, \Gamma_n, T_n)$ will obey the inequalities

$$d(\Gamma_n) < 2^{-2n}, \quad \|S_n - \lambda e_0 - \mu e_1\| < 2^{-2n}, \quad n \in \mathbf{N}. \tag{2}$$

Let $\mathcal{A}_n$ denote the set of those segments Δ_k in Γ_n for which $t_k \notin (\bigcup_n E_n) \cup (\bigcup_n G_n)$. From the definition of the norm in the space $l_1[0, 1]$ it readily follows that

$$2^{-2n} \geq \|S_n - \lambda e_0 - \mu e_1\| \geq \left\| \sum_{\Delta_k \in \mathcal{A}_n} e_{t_k} \Delta_k \right\| = \sum_{\Delta_k \in \mathcal{A}_n} |\Delta_k|.$$

Hence, $\sum_{\Delta_k \in \mathcal{A}_n} |\Delta_k| \to 0$ when $n \to \infty$. Therefore, if we discard from the sum S_n all the terms corresponding to the segments $\Delta_k \in \mathcal{A}_n$, then all the limit relations that we require to be satisfied for the sequence of integral sums are preserved. Further, if from the integral sum S_n we discard all the terms corresponding to the segments that contain points from E_k or G_k with $k \leq n$, then the sum does not change by more than $M(f)2^{n+1}d(\Gamma_n) < 3 \cdot 2^{-n+1}$. Hence, the described procedure will not affect the limit relations.

A segment in the partition Γ_n will be said to be "*squeezed*" if it lies between points of the form

$$a = \frac{j}{2^{n+k-1}} - \frac{1}{4^{n+k}} \quad \text{and} \quad b = \frac{j}{2^{n+k-1}} + \frac{1}{4^{n+k}}$$

for some j and k. The total length of all squeezed segments in Γ_n does not exceed 2^{-n+2}. For this reason from this point on we shall assume that the integral sum does not contain terms corresponding to squeezed segments.

After all segments listed above are discarded, inequality (2) may deteriorate and become

$$\|S_n - \lambda e_0 - \mu e_1\| \leq \varepsilon_n^2, \tag{3}$$

where ε_n is some sequence that converges to 0.

Denote by $C_{n,j}$ [resp. $D_{n,j}$] the set of indices k for which the points t_k in the collection T_n belong to E_j [resp. G_j]. Also, set

$$C_n = \bigcup_{j=n+1}^{\infty} C_{n,j}, \quad D_n = \bigcup_{j=n+1}^{\infty} D_{n,j}.$$

Since the "superfluous segments" were discarded from the integral sum, $S_n = \sum_{k \in C_n} f(t_k)|\Delta_k| + \sum_{k \in D_n} f(t_k)|\Delta_k|$. Writing inequality (3) in more detail and observing that the expressions of the form $f(t_k) - e_0$ with $k \in C_n$ and $f(t_k) - e_1$ with $k \in D_n$ contain the unit vectors e_0 and e_1 with the coefficient zero and that for $k \in D_n$ and $k \in C_n$ these expressions contain different unit vectors, we obtain

$$\varepsilon_n^2 \geq \|S_n - \lambda e_0 - \mu e_1\| = \Bigg\| e_0 \Bigg(\sum_{k \in C_n} |\Delta_k| - \lambda \Bigg) + e_1 \Bigg(\sum_{k \in D_n} |\Delta_k| - \mu \Bigg)$$

$$+ \sum_{k \in C_n} [f(t_k) - e_0]|\Delta_k| + \sum_{k \in D_n} [f(t_k) - e_1]|\Delta_k| \Bigg\|$$

$$= \Bigg| \lambda - \sum_{k \in C_n} |\Delta_k| \Bigg| + \Bigg| \mu - \sum_{k \in D_n} |\Delta_k| \Bigg|$$

$$+ \Bigg\| \sum_{k \in C_n} [f(t_k) - e_0]|\Delta_k| \Bigg\| + \Bigg\| \sum_{k \in C_n} [f(t_k) - e_0]|\Delta_k| \Bigg\|.$$

Consequently,

$$\left\|\sum_{k\in C_n}[f(t_k)-e_0]|\Delta_k|\right\|+\left\|\sum_{k\in D_n}[f(t_k)-e_1]|\Delta_k|\right\|\le\varepsilon_n^2 \tag{4}$$

and

$$\left|\lambda-\sum_{k\in C_n}|\Delta_k|\right|+\left|\mu-\sum_{k\in D_n}|\Delta_k|\right|\le\varepsilon_n^2. \tag{5}$$

Since the segments Δ_k with $k\in C_n$ and $k\in D_n$ form almost the entire partition Γ_n, inequality (5) implies that

$$\lambda+\mu=\lim_{n\to\infty}\sum_{k\in C_n}|\Delta_k|+\lim_{n\to\infty}\sum_{k\in D_n}|\Delta_k|=1.$$

It remains to show that λ and μ cannot be simultaneously different from zero. To this end we need to examine more closely the structure of the sets C_n and D_n.

Let us fix n for the moment and define the numbers $R(j)$ as the difference between 2^j and the number of elements in the set $C_{n,j}$. Let us write in more detail the first term of inequality (4):

$$\left\|\sum_{k\in C_n}[f(t_k)-e_0]|\Delta_k|\right\|=\sum_{j=n+1}^{\infty}\left\|\sum_{k\in C_{n,j}}[f(t_k)-e_0]|\Delta_k|\right\|$$

$$=\sum_{j=n+1}^{\infty}\left\|\sum_{k\in C_{n,j}}\left(\frac{1}{2^j}\sum_{\alpha\in E_j}e_\alpha-e_{t_k}\right)|\Delta_k|\right\|$$

$$=\sum_{j=n+1}^{\infty}\left[\sum_{k\in C_{n,j}}\left|\frac{1}{2^j}\sum_{i\in C_{n,j}}|\Delta_i|-|\Delta_k|\right|+R(j)\frac{1}{2^j}\sum_{i\in C_{n,j}}|\Delta_i|\right].$$

Combining this with inequality (4), we obtain

$$\sum_j\left[\frac{R(j)}{2^j}\sum_{i\in C_{n,j}}|\Delta_i|\right]\le\varepsilon_n^2.$$

Let U denote the set of indices j for which $R(j)<\varepsilon_n 2^j$. By the last inequality,

$$\varepsilon_n^2\ge\sum_{j\notin U}\left[\frac{R(j)}{2^j}\sum_{i\in C_{n,j}}|\Delta_i|\right]\ge\varepsilon_n\sum_{j\notin U}\sum_{i\in C_{n,j}}|\Delta_i|,$$

which means that the set U is "big" in the sense that

$$\sum_{j \notin U} \sum_{i \in C_{n,j}} |\Delta_i| \leq \varepsilon_n. \tag{6}$$

Since the integral sum involves only a finite number of segments, U is a finite set. Let $j_0 = \max\{j : j \in U\}$.

Let us show that the set $U \setminus \{j_0\}$ is "small" in a sense analogous to inequality (6). To this end let us recall how the set E_n was constructed: it consists of 2^m points of the segment $[0, 1]$, and to the left and to the right of each point of the form $k/2^{m-1}$ there is a point from E_m at the distance $8^{-m}\sqrt{2}$. That is to say, each point of the form $k/2^{m-1}$ is "squeezed" between two points of E_m. Let $j \in U \setminus \{j_0\}$. Then $C_{n,j}$ consists of $2^j - R(j) \geq 2^j(1 - \varepsilon_n)$ elements and hence it "squeezes" between the points t_k with $k \in C_{n,j}$ at least $2^{j-1}(1 - 2\varepsilon_n)$ points of the form $k/2^{n-1}$. Since all the "squeezed" segments were discarded from our integral sum, $R(j_0) \geq 2^j(1-2\varepsilon_n)$. Further, since $j_0 \in U$, combining the inequality $R(j_0) < \varepsilon_n 2^{j_0}$ from the definition of the set U and the last inequality, we conclude that

$$2^{j_0} \frac{\varepsilon_n}{1 - 2\varepsilon_n} \geq 2^j, \quad j \in U \setminus \{j_0\}. \tag{7}$$

Now let us estimate the sum $\sum_{j \in U \setminus \{j_0\}} \sum_{k \in C_{n,j}} |\Delta_k|$. To this end we shall break it into two sums, denoted Σ_1 and Σ_2: the first [resp. second] involves the segments of length smaller [resp. larger] than 2^{-j_0+2}. Then inequality (7) yields

$$\Sigma_1 < \sum_{j \in U \setminus \{j_0\}} \sum_{k \in C_{n,j}} 2^{-j_0+2} \leq 2 \cdot 2^{j_0} \frac{\varepsilon_n}{1 - 2\varepsilon_n} \cdot 2^{-j_0+2} < 10\varepsilon_n.$$

Further, each segment Δ_k from Σ_2 contains at least $|\Delta_k| \cdot 2^{j_0-2}$ points of the set E_{j_0}. Consequently, $R(j_0) \geq 2^{j_0-2}\Sigma_2$, and since $j_0 \in U$, $\Sigma_2 < 4\varepsilon_n$. Hence,

$$\sum_{j \in U \setminus \{j_0\}} \sum_{k \in C_{n,j}} |\Delta_k| = \Sigma_1 + \Sigma_2 < 14\varepsilon_n. \tag{8}$$

Combining the inequalities (5), (6), and (8), we conclude that $\sum_{k \in C_{n,j_0}} |\Delta_k|$ is almost equal to λ. In much the same way one can show that there exists an index j_1 for which the sum $\sum_{k \in D_{n,j_1}} |\Delta_k|$ is almost equal to μ.

Suppose $j_0 \geq j_1$. Then, using estimates similar to those used in the proof of inequality (8), we conclude that the quantity $\sum_{k \in D_{n,j_1}} |\Delta_k|$ is small. If $j_0 < j_1$, then $\sum_{k \in C_{n,j_0}} |\Delta_k|$ is small. This means that either λ or μ is estimated from above by a quantity that tends to zero when $n \to \infty$. Therefore, $\lambda e_0 + \mu e_1$ is equal to either e_0 or e_1. □

REMARK 1. By modifying the foregoing arguments one can readily produce a function f such that $\mathrm{I}(f)$ consists of any given finite number of points. In this connection one can formulate the following

QUESTION. Can any bounded closed subset of $l_1[0, 1]$ serve as the set $\mathrm{I}(f)$ for some function?

§3. Separability of the Space and the Structure of I(f)

The example constructed in the preceding section, of a function f for which I(f) consists of two points, lives in a nonseparable Hilbert space. It is natural to ask whether the construction can be carried over to some simpler, separable space. Let us try to achieve this, modifying the first row of the definition of $f(t)$ in Example 2: for $t \notin (\bigcup_n E_n) \cup (\bigcup_n G_n)$ take $f(t)$ equal to some fixed basis vector, say $e_{1/2}$. The rest of the construction remains essentially the same, the resulting function takes values in a separable space, and I(f) remains nonconvex: as before, a point of the form $\lambda e_0 + \mu e_1$ can belong to I(f) only if $\lambda = 1$, $\mu = 0$, or $\lambda = 0$, $\mu = 1$. However, I(f) is no longer a two-point set: it contains "parasitic elements" of the form $\lambda e_0 + (1-\lambda)e_{1/2}$ and $\lambda e_1 + (1-\lambda)e_{1/2}$, $0 \le \lambda \le 1$. Below it will be shown that this complication is not a drawback of the construction used, but rather a general feature for functions with values in a separable space. More precisely, we will prove that in the separable case the set I(f) is star-shaped (i.e., there exists an $x \in \mathrm{I}(f)$ such that $\lambda x + (1-\lambda)y \in \mathrm{I}(f)$ for all $y \in \mathrm{I}(f)$ and all $\lambda \in [0,1]$). This will also show that in this case I(f) is not empty.

A partition $\tilde{\Gamma}$ is said to be a *refinement* of the partition Γ if all the partition points of Γ are partition points for $\tilde{\Gamma}$.

DEFINITION 2. An integral sum $S(f,\Gamma,T)$ is said to be ε-*mixed* if, for any refinement $\tilde{\Gamma}$ of the partition Γ, any choice $\tilde{T}$ of points $\tilde{t}_i$, and any $\lambda \in [0,1]$ one can find a third partition, Γ_λ, and a collection of points T_λ, such that $d(\Gamma_\lambda) \le d(\Gamma)$ and

$$\|S(f,\Gamma_\lambda,T_\lambda) - [\lambda S(f,\Gamma,T) + (1-\lambda)S(f,\tilde{\Gamma},\tilde{T})]\| < \varepsilon.$$

In what follows we shall denote by μ^* [resp. μ_*] the outer [resp. inner] Lebesgue measure on the segment $[0,1]$.

LEMMA 1. *Let X be a separable Banach space, $f\colon [0,1] \to X$ a bounded function. Then for any $\varepsilon > 0$ the function f has ε-mixed integral sums.*

PROOF. All values of f lie in the ball B of radius $M = M(f)$ centered at zero. Partition B into disjoint sets $\{B_i\}_{i=1}^\infty$ of diameters smaller than or equal to $\varepsilon/4$ (this means that if $x, y \in B_i$ for some i, then $\|x-y\| \le \varepsilon/4$). The existence of such a partition of B is guaranteed by the separability of the space X. Consider the preimages $f^{-1}(B_i)$. This is a system of disjoint, generally speaking nonmeasurable sets, which cover the segment $[0,1]$. Choose a number n such that the outer measure of the set $A = \bigcup_{k=1}^n f^{-1}(B_k)$ is larger than $1 - \varepsilon/(8Mn)$. Cover $A_1 = f^{-1}(B_1)$ by a countable set of disjoint segments whose total length equals $\mu^*(A_1) + \varepsilon/(8Mn)$. Choose among these a finite number of segments $\Delta_1, \dots, \Delta_{m_1}$ whose total length $\sum_{i=1}^{m_1} |\Delta_i|$ is larger than $\mu^*(A_1)$. Denote the part of the set $f^{-1}(B_2)$ that is not contained in $\bigcup_{k=1}^{m_1} \Delta_k$ by A_2. Clearly, $f(A_2) \subset B_2$. Proceeding with A_2 in the same manner we did with A_1, select segments $\{\Delta_i\}_{i=m_1+1}^{m_2}$ that cover the main part of A_2:

$$\mu^*(A_2) < \sum_{i=m+1+1}^{m_2} |\Delta_i| < \mu^*(A_2) + \frac{\varepsilon}{8Mn}, \qquad \mu^*\left(A_2 \setminus \bigcup_{i=m_1+1}^{m_2} \Delta_i\right) < \frac{\varepsilon}{8Mn}.$$

Make this selection so that all segments Δ_i with $i \in [1, m_2]$ are disjoint. Repeat inductively this process n times, where at the jth step the set A_j is defined as $f^{-1}(B_j) \setminus \bigcup_{k=1}^{m_{j-1}} \Delta_k$. This yields a set of segments $\{\Delta_k\}_{k=1}^{m_n}$, of total length smaller that $1 - \varepsilon/(8M)$, such that for each k there exists a number $j(k) \in 1, \dots, n$ for which

$$\sum_{k=1}^{m_n} \mu_* \left(\Delta_k \setminus f^{-1}(B_{j(k)})\right) < \frac{\varepsilon}{8M}.$$

Next, adjoin to $\{\Delta_k\}_{k=1}^{m_n}$ a finite number of segments $\{\Delta_k\}_{k=m_n+1}^{N}$ so that the system $\{\Delta_k\}_{k=1}^{N}$ will be a partition of $[0,1]$. Then $\sum_{k=m_n+1}^{N} |\Delta_k| \le \varepsilon/(8M)$. Choose $j(k)$ for $k > m_n$ in arbitrary manner. Then

$$\sum_{k=1}^{N} \mu_* \left(\Delta_k \setminus f^{-1}(B_{j(k)})\right) < \frac{\varepsilon}{4M}. \tag{9}$$

Denote the partition formed by the segments Δ_k by Γ. Now pick a collection T of points $t_k \in \Delta_k$ as follows: if $\Delta_k \cap f^{-1}(B_{j(k)}) \neq \emptyset$, take $t_k \in f^{-1}(B_{j(k)})$, while if the above intersection is empty take for t_k any point in Δ_k.

We claim that $S(f, \Gamma, T)$ is an ε-mixed integral sum. Indeed, let $\tilde{\Gamma}$ be a partition of $[0,1]$ into segments C_i, with $\tilde{\Gamma}$ a refinement of Γ, and let $\tilde{T}$ be an arbitrary collection of points $\tilde{t}_i \in C_i$. Then $S(f, \tilde{\Gamma}, \tilde{T}) = \sum f(\tilde{t}_i)|C_i|$. Let us construct the needed partition Γ_λ. To this end we replace each segment C_i by a segment $C_i' \subset C_i$ such that $|C_i'| = (1-\lambda)|C_i|$ and $\tilde{t}_i \in C_i'$. Each set $\Delta_k \setminus \bigcup_i C_i'$ consists of a finite number of segments R_j. Now let Γ_λ be the partition consisting of the segments C_i' and R_j. Obviously,

$$\sum_{C_i' \subset \Delta_k} |C_i'| = (1-\lambda)|\Delta_k|, \qquad \sum_{R_j \subset \Delta_k} |R_j| = \lambda |\Delta_k|.$$

The needed collection of points T_λ is selected as follows. In each segment C_i' take the same point $\tilde{t}_i$ that we took in C_i. Let U_k denote the set of indices j for which $R_j \subset \Delta_k$ and $R_j \cap f^{-1}(B_{j(k)}) \neq \emptyset$. For $j \in U_k$ pick a point $t_j^* \in R_j$ that also lies in $f^{-1}(B_{j(k)})$, whereas for $j \notin U_k$ pick an arbitrary $t_j^* \in R_j$. By (9),

$$\sum_{j \notin \bigcup_{k=1}^{N} U_k} |R_j| \le \frac{\varepsilon}{4M}, \qquad \sum_{k=1}^{N} \left| \lambda |\Delta_k| - \sum_{j \in U_k} |R_j| \right| \le \frac{\varepsilon}{4M}. \tag{10}$$

Now let us write the constructed integral sum in detail:

$$S(f, \Gamma_\lambda, T_\lambda) = \sum f(\tilde{t}_i)|C_i'| + \sum_{k=1}^{N} \sum_{j \in U_k} f(t_j^*)|R_j| + \sum_{j \notin \bigcup_{k=1}^{N} U_k} f(t_j^*)|R_j| =$$

$$= (1-\lambda)S(f, \tilde{\Gamma}, \tilde{T}) + \sum_{k=1}^{N} \sum_{j \in U_k} f(t_j^*)|R_j| + \sum_{j \notin \bigcup_{k=1}^{N} U_k} f(t_j^*)|R_j|.$$

By inequality (10), we have

$$\|S(f, \Gamma_\lambda, T_\lambda) - \lambda S(f, \Gamma, T) - (1-\lambda)S(f, \tilde{\Gamma}, \tilde{T})\|$$

$$\leq \left\| \sum_{k=1}^{N} \left(\sum_{j \in U_k} f(t_j^*)|R_j| - \lambda f(t_k)|\Delta_k| \right) \right\| + M\frac{\varepsilon}{4M}$$

$$= \left\| \sum_{k=1}^{N} \left[\sum_{j \in U_k} \left(f(t_j^*) - f(t_k)\right)|R_j| + f(t_k)\left(\sum_{j \in U_k} |R_j| - \lambda|\Delta_k| \right) \right] \right\| + \frac{\varepsilon}{4}$$

$$\leq \left\| \sum_{k=1}^{N} \left[\sum_{j \in U_k} \left(f(t_j^*) - f(t_k)\right)|R_j| \right] \right\| + \frac{\varepsilon}{2}.$$

Since $f(t_k)$ and $f(t_j^*)$ lie in $B_{j(k)}$ whenever $j \in U_k$, and since the diameter of $B_{j(k)}$ is not larger than $\varepsilon/4$, one has the bound

$$\left\| \sum_{k=1}^{N} \left[\sum_{j \in U_k} \left(f(t_j^*) - f(t_k)\right)|R_j| \right] \right\| \leq \varepsilon/4,$$

whence

$$\|S(f, \Gamma_\lambda, T_\lambda) - \lambda S(f, \Gamma, T) - (1-\lambda)S(f, \tilde{\Gamma}, \tilde{T})\| < \varepsilon,$$

as claimed. □

REMARK 2. The above construction shows that the partition Γ in Lemma 1 can be taken as fine as one wishes.

LEMMA 2. *Let X be a separable space, $f\colon [0,1] \to X$ a bounded function. Then there exists an element $x \in \mathrm{I}(f)$ such that, for any $\varepsilon > 0$, x be arbitrarily well approximated by ε-mixing integral sums of f that correspond to arbitrarily fine partitions of the segment $[0,1]$.*

PROOF. Set $M = M(f)$ and pick a sequence $\varepsilon_n \searrow 0$ such that $\sum_{n=1}^{\infty} \varepsilon_n < \infty$. We will construct by induction a sequence of integral sums $S_n = S(f, \Gamma_n, T_n)$ with the following properties: (1) S_n is an ε_n-mixed integral sum with $d(\Gamma_n) < \varepsilon_n$, and (2) $\|S_{n+1} - S_n\| \leq \varepsilon_n$ for all n. Once this construction is successfully carried out, condition (2) guarantees that the sequence S_n converges to some $x \in X$, and since $\lim_{n\to\infty} d(\Gamma_n) = 0$, it follows that $x \in \mathrm{I}(f)$.

Now to the induction process. Construct the sum $S_1 = S(f, \Gamma_1, T_1)$ by repeating the arguments of Lemma 1 with $\varepsilon = \varepsilon_1$, taking care that the inequality $d(\Gamma_1) < \varepsilon_1$ will be satisfied (which is possible according to Remark 2). Then the sum S_1 will be ε_1-mixing. Let N, B_i, Δ_k, and $j(k)$ be as in the proof of Lemma 1. Then (9) holds with $\varepsilon = \varepsilon_1$. To obtain the partition Γ_2, we proceed by analogy with Lemma 1 for each individual segment Δ_k. Thus, partition each set B_i into a countable number of subsets of diameters no larger than $\varepsilon_2/4$. Reindex these sets and denote them by B_i'. Choose the new partition Γ_2 of $[0,1]$, consisting of the segments $\{\Delta_i'\}_{i=1}^{N'}$, so that it will enjoy the following property: for any index i one can find an index $j'(i)$ such that

$$\sum_{i=1}^{N'} \mu_* \left(\Delta_i' \setminus f^{-1}(B_{j'(i)}')\right) < \frac{\varepsilon_2}{4M}, \tag{11}$$

and for all indices i for which $\Delta_i' \subset \Delta_k$ the set $B_{j'(i)}'$ is "almost always" a subset of $B_{j(k)}$. Here "almost always" is understood in the following sense: if H_k denotes the set of all i for which $\Delta_i' \subset \Delta_k$, but $B_{j'(k)}' \not\subset B_{j(k)}$, then

$$\sum_{k=1}^{N} \sum_{i \in H_k} |\Delta_i'| < \frac{\varepsilon_1}{4M}. \tag{12}$$

That this can be achieved follows from inequality (9). Now choose the collection T_2 of points t_i' so that whenever $\Delta_i' \cap f^{-1}(B_{j'(i)}') \neq \emptyset$ the point t_i' belongs to $f^{-1}(B_{j'(i)}')$.

Condition (12) and the fact that $d(B_{j(k)}) < \varepsilon_1/4$ yield

$$\|S(f, \Gamma_1, T_1) - S(f, \Gamma_2, T_2)\| < \varepsilon_1,$$

and inequality (11) shows that $S_2 = S(f, \Gamma_2, T_2)$ is an ε_2-mixed integral sum.

Next, carrying out the construction of Lemma 1 on each segment of the partition Γ_2, we produce an ε_3-mixed integral sum S_3 such that $\|S_3 - S_2\| \leq \varepsilon_2$. Continuing in this manner, we obtain the desired sequence $\{S_n\}$. □

Notice that the last lemma shows, in particular, that for functions with values in a separable space the set $\mathrm{I}(f)$ is nonempty. This result was first obtained by H. W. Ellis [1]. Repeating the construction of Example 1 for the space $l_1[0,1]$, it can be readily verified that for arbitrary nonseparable spaces Ellis' theorem fails. Nevertheless, as the results of the following section imply, for *certain* nonseparable spaces (for instance, nonseparable L_p spaces) Ellis' theorem does remain valid. In this connection there arises the following question, which is still open: How to characterize the spaces for which the assertion of Ellis' theorem holds true?

THEOREM 3 (V. M. KADETS [7]). *Let f be a bounded function on the segment $[0, 1]$ with values in a separable space X. Then $\mathrm{I}(f)$ is a star-shaped set.*

PROOF. Let $x \in \mathrm{I}(f)$ be the element provided by Lemma 2, i.e., x can be approximated by ε-mixing integral sums of f. We claim that $\mathrm{I}(f)$ is star-shaped with respect to x. Indeed, let $y \in \mathrm{I}(f)$ and $\lambda \in [0, 1]$. We need to show that $\lambda x + (1 - \lambda)y \in \mathrm{I}(f)$. Take an arbitrary $\varepsilon > 0$ and pick an $(\varepsilon/4)$-mixed integral sum $S_x = S(f, \Gamma, T)$ such that $d(\Gamma) < \varepsilon$ and $\|x - S_x\| \leq \varepsilon/4$. Let n be the number of segments in the partition Γ. Pick an integral sum $S_y = S(f, \Gamma_1, T_1)$ such that $d(\Gamma_1) < \varepsilon/(8nM(f))$ and $\|y - S_y\| \leq \varepsilon/4$. Add to Γ_1 all the endpoints of the segments of partition Γ and denote the resulting partition by $\tilde{\Gamma}$. On the segments of $\tilde{\Gamma}$ that coincide with segments of Γ_1 preserve the choice of the points $t_i \in T_1$, while on the remaining segments choose the points in arbitrary manner. Denote the resulting collection of points by $\tilde{T}$. The total length of the segments from Γ_1 that contain no endpoints of segments from Γ is no larger than $nd(\Gamma_1) < \varepsilon/(8M(f))$. Hence, the integral sum $\tilde{S}_y = S(f, \tilde{\Gamma}, \tilde{T})$ satisfies $\|S_y - \tilde{S}_y\| \leq 2M(f) \cdot \varepsilon/(8M(f)) = \varepsilon/4$ and $\|y - \tilde{S}_y\| \leq \varepsilon/2$. Moreover, $\tilde{\Gamma}$ is a refinement of Γ. Therefore, since the sum S_x is $(\varepsilon/4)$-mixed, there exist a partition Γ_λ and a collection of points T_λ such that $d(\Gamma_\lambda) \leq d(\Gamma) \leq \varepsilon$ and $\|S(f, \Gamma_\lambda, T_\lambda) - \lambda S_x - (1 - \lambda)\tilde{S}_y\| \leq \varepsilon/4$. Thus, we have produced an integral sum $S_\lambda = S(f, \Gamma_\lambda, T_\lambda)$ with the property that

$$\begin{aligned} \|S_\lambda - \lambda x - (1 - \lambda)y\| &\leq \|S_\lambda - \lambda S_x - (1 - \lambda)\tilde{S}_y\| + \lambda\|x - S_x\| + (1 - \lambda)\|y - \tilde{S}_y\| \\ &\leq \frac{\varepsilon}{4} + \frac{\varepsilon}{4} + \frac{\varepsilon}{2} = \varepsilon. \end{aligned}$$

Thanks to the arbitrariness of ε, this means that $\lambda x + (1 - \lambda)y$ belongs to $\mathrm{I}(f)$, as claimed. □

§4. Connection with the Weak Topology

DEFINITION 3. We say that Banach space X *belongs to the convex class* (and write $X \in$ **CONV**) if for any bounded function $f \colon [0, 1] \to X$ the set $\mathrm{I}(f)$ is convex.

In §1 we have shown that the B-convex spaces belong to **CONV**, whereas, as it became clear in §2 and the beginning of §3, the space l_1, and hence all the spaces that contain an isomorphic copy of l_1, do not belong to **CONV**. One is led to the problem of giving a full characterization of the class **CONV**. This problem turned out to be quite challenging, and a complete solution is not yet available. Moreover, we do not know whether such a classical space as c_0 belongs to **CONV**. However, substantial progress on the path to a full description of the class **CONV** can be achieved by bringing into consideration weakly convergent sequences of integral sums.

DEFINITION 4. The *set of weak limits of Riemann integral sums* of a bounded function $f \colon [0, 1] \to X$ is the set $\mathrm{WI}(f)$ of all points of X that are limits of weakly convergent sequences of integral sums corresponding to finer and finer partitions.

DEFINITION 5. We say that a space X *belongs to the weakly convex class* (and write $X \in \mathbf{WCONV}$) if the set $\mathrm{WI}(f)$ is convex for any bounded function $f\colon [0,1] \to X$.

Obviously, $\mathrm{I}(f) \subset \mathrm{WI}(f)$ for any function f. In this section we will show that actually for many spaces X the sets $\mathrm{I}(f)$ and $\mathrm{WI}(f)$ coincide for any function $f\colon [0,1] \to X$ with $M(f) < \infty$. We call this property of a space the *coincidence property*, and use the notation $X \in \mathbf{COIN}$. Using the complete description of the class **WCONV** that will be given below and the coincidence property, we shall prove that $X \in \mathbf{CONV}$ for a wide class of spaces X. All the results presented here belong the V. M. Kadets [5, 6].

We next turn to the investigation of the structure of the set $\mathrm{WI}(f)$. We begin by listing two elementary properties: if A is a convex set and $f\colon [0,1] \to A$, then $\mathrm{WI}(f) \subset A$; if X is reflexive and the function $f\colon [0,1] \to X$ is bounded, then $\mathrm{WI}(f) \neq \emptyset$.

A function f will be said to be *weakly Riemann-integrable* if any sequence of integral sums of f corresponding to increasingly finer partitions is weakly convergent. It is obvious that for such a function f the set $\mathrm{WI}(f)$ reduces to a single point.

THEOREM 4. *In order that the Banach space X be reflexive it is necessary and sufficient that any bounded function $f\colon [0,1] \to X$ for which* $\mathrm{WI}(f)$ *consists of a single point be weakly Riemann-integrable.*

PROOF. NECESSITY. Suppose that X is reflexive and $\mathrm{WI}(f)$ consists of a single point x. We need to show that any sequence $S_n = S(f, \Gamma_n, T_n)$ such that $d(\Gamma_n) \to 0$ converges weakly to x. Assume the contrary, i.e., there is a functional $x^* \in X^*$ such that the sequence $x^*(S_n)$ does not converge to $x^*(x)$. Since X is reflexive and the sequence $\{S_n\}$ is bounded, one can find a subsequence $\{S'_k\} \subset \{S_n\}$ such that $S'_k \to y$ weakly and $\lim_{k\to\infty} x^*(S'_k) \neq x^*(x)$. But then $y \neq x$ and $y \in \mathrm{WI}(f)$, which contradicts the assumption on $\mathrm{WI}(f)$.

SUFFICIENCY. Suppose X is not reflexive. Then according to a result in [9], one can choose a basic sequence $\{e_k\}_{k=1}^{\infty}$ of norm-one elements of X that is separated from zero by a common linear functional $x^* \in X^*$: $\inf\{x^*(e_k)\} = a > 0$. Define the function $f\colon [0,1] \to X$ as follows: $f(1/2) = e_1$, $f(1/4) = f(3/4) = e_2$, $f(1/8) = f(3/8) = f(5/8) = f(7/8) = e_3$, … , $f((2k-1)/2^n) = e_n$ for $k = 1, 2, \ldots, 2^{n-1}$, $n \in \mathbf{N}$, and $f(t) = 0$ for the remaining values of t. Suppose that some sequence $S_n = S(f, \Gamma_n, T_n)$ of integral sums of f for which $d(\Gamma_n) \to 0$ converges weakly to some element $x = \sum_{k=1}^{\infty} a_k e_k$. Since for each element e_k the coefficient with which it may appear in S_n is not larger than $2^k d(\Gamma_n)$, we see that $a_k = 0$ for all k. Therefore, $\mathrm{WI}(f) = \{0\}$. On the other hand, if we take the partition Γ'_n obtained by dividing $[0,1]$ into 2^{n-1} equal parts and the collection T'_n consisting of the points $t_k = (2k-1)/2^n$, then $S(f, \Gamma'_n, T'_n) = e_n$. Since the sequence $\{e_n\}$ does not coverge weakly to zero, the constructed function f is not weakly Riemann-integrable, which completes the proof of the theorem. □

THEOREM 5. *The following two conditions are equivalent:* 1) X *does not contain an isomorphic copy of* l_1; 2) $X \in$ **WCONV**.

PROOF. As we have observed earlier, modifying slightly the construction of Example 2 one can built a bounded function $g\colon [0,1] \to l_1$ for which $\mathrm{I}(g)$ is not convex. In l_1 every weakly convergent sequence converges strongly, and so $\mathrm{WI}(g) = \mathrm{I}(g)$. Therefore $\mathrm{WI}(g)$ is not convex and $l_1 \notin$ **WCONV**. Since the property **WCONV** is obviously inherited by subspaces, from the assumption that $X \in$ **WCONV** it follows that X does not contain isomorphic copies of l_1.

Now let us prove the implication 1) $\Rightarrow$ 2). Let $f\colon [0,1] \to X$ be a bounded function and $x, y \in \mathrm{WI}(f)$. Pick an arbitary $\lambda \in [0,1]$. We need to show that $\lambda x + (1-\lambda)y \in \mathrm{WI}(f)$. Suppose $S(f, \Gamma_n, T_n) \to x$ weakly, $d(\Gamma_n) \to 0$, $S(f, \Gamma'_n, T'_n) \to x$ weakly, $d(\Gamma'_n) \to 0$. Further, let $\{x_i^*\}_{i=1}^m$ be arbitrary elements of X^*. Define the operator $U\colon X \to l_\infty^{(m)}$ by the rule $Ux = (x_1^*(x), \dots, x_m^*(x))$. Consider the auxiliary function $\overline{f}\colon [0,1] \to l_\infty^{(m)}$ given by $\overline{f}(t) = Uf(t)$. Since the operator U is continuous, Ux and Uy belong to $\mathrm{I}(\overline{f})$. Next, since the space $l_\infty^{(m)}$ is finite-dimensional, the set $\mathrm{I}(\overline{f})$ is convex. Consequently, there exist a partition Γ''_n and a collection of points T''_n such that $S(\overline{f}, \Gamma''_n, T''_n) \to U(\lambda x + (1-\lambda)y)$, $d(\Gamma''_n) \to 0$, and $T''_n \subset \bigcup_{n=1}^\infty (T_n \cup T_n)$ (that the last condition can be satisfied follows from an examination of the proof of Theorem 1). Recalling the definitions of U and of the norm in $l_\infty^{(m)}$, we obtain

$$\lim_{n\to\infty} \left(\max_{1\le i\le m} |x_i^* (\lambda x + (1-\lambda y) - S(f, \Gamma''_n, T''_n))| \right) + d(\Gamma''_n) = 0. \tag{13}$$

Consider the auxiliary Banach space $\overline{X}$ consisting of the pairs (x, d), where $x \in X$, $d \in \mathbf{R}$, with the norm $\|(x,d)\| = \|x\| + |d|$. Thanks to the arbitrariness of the functionals x_i^*, relation (13) means that $(\lambda x + (1-\lambda)y, 0)$ is a limit point in the weak topology for the set $V \subset \overline{X}$ of the points of the form $(S(f, \Gamma, T), d(\Gamma))$ with $T \subset \bigcup_{k=1}^\infty (T_n \cup T'_n)$. The last condition ensures that the set V is separable. By a theorem of H. P. Rosethal [13], in a space that does not contain a copy of l_1, the weak closure and the weak sequential closure of a bounded separable set coincide. Hence, there exists a pair Γ'''_n, T'''_n for which the sequence $(S(f, \Gamma'''_n, T'''_n), d(\Gamma'''_n))$ converges weakly to $(\lambda x + (1-\lambda)y, 0)$ in $\overline{X}$. But this means that $S(f, \Gamma'''_n, T'''_n) \to \lambda x + (1-\lambda)y$ weakly and $d(\Gamma'''_n) \to 0$, i.e., $\lambda x + (1-\lambda)y \in \mathrm{WI}(f)$, as needed. □

Thus, we have obtained a complete description of the class **WCONV**. Now our task is to study the coincidence property. Clearly, spaces which, like l_1, have the Schur property (i.e., in which weak and strong convergence of sequences coincide), belong to the class **COIN**. Our immediate objective is to show that the spaces that are the "farthest" from l_1, namely, the B-convex spaces, also belong to **COIN**. The following assertion, which is of independent interest as well, will play here an auxiliary role.

LEMMA 3. *Suppose the Banach space* X *has infratype* $p > 1$ *with constant* $C > 0$. *Let* $\{A_i\}_{i=1}^n$ *be an arbitrary finite collection of bounded subsets of* X, $d_i =$

diam A_i, $i = 1, \dots, n$, and let $\{b_i\}_{i=1}^n$ be points such that $b_i \in \operatorname{conv} A_i$, $i = 1, \dots, n$. Then one can choose points $a_i \in A_i$, $i = 1, \dots n$, such that $\|\sum_{i=1}^n (a_i - b_i)\| \le C_1 \left(\sum_{i=1}^n d_i^p\right)^{1/p}$, where $C_1 = 2C\sum_{k=1}^\infty 2^{(\frac{1}{p}-1)k}$.

PROOF. Using the assumption that $b_i \in \operatorname{conv} A_i$, let us choose numbers $0 = m_0 < m_1 < m_2 < \cdots < m_n$, elements $\{x_k\}_{k=1}^{m_n}$, and numbers $\{\lambda_k\}_{k=1}^{m_n}$, with the following properties: $x_k \in A_i$ for $m_{i-1} < k \le m_i$; $\lambda_k > 0$ for all k;

$$\sum_{k=m_i+1}^{m_{i+1}} \lambda_k = 1; \qquad \sum_{k=m_i+1}^{m_{i+1}} \lambda_k x_k = b_i, \quad i = 0, 1, \dots, n-1.$$

Our proof will rely on the same idea (borrowed from the paper [11]) as the proof of Lemma 2.3.3. Specifically, we shall assume that all λ_k are finite dyadic fractions (this assumption does not result in a loss of generality, since it can be realized by slightly perturbing the numbers λ_k). Let N be the largest length of the dyadic decomposition of the numbers λ_k. Then each λ_k admits can be written as $\lambda_k = \sum_{l=0}^N \varepsilon_{k,l} 2^{-l}$, where the $\varepsilon_{k,l}$ are equal to 0 or 1.

Let $k(1) < k(2) < \cdots < k(r)$ be the numbers for which $\varepsilon_{k(l),N} = 1$. Since $\sum_{k=m_i+1}^{m_{i+1}} \lambda_k = 1$, the number r_i of indices $k(l)$ for which $m_i < k(l) \le m_{i+1}$ is even for any i. For the same reason, $r_i \le 2^N$ for all $i < n$. Introduce the auxiliary elements $g_l = x_{k(2l-1)} - x_{k(2l)}$, $l = 1, \dots, r/2$. Clearly, $\|g_l\| \le d_{i+1}$ for $m_i < k(2l) < m_{i+1}$. Since X has infratype p, we can choose signs $\alpha_l = \pm 1$ such that

$$\left\|\sum_{l=1}^{r/2} \alpha_l g_l\right\| \le C\left(\sum_{l=1}^{r/2} \|g_l\|^p\right)^{1/p}. \tag{14}$$

Now introduce the new numbers $\{\lambda_{k,1}\}_{k=1}^{m_n}$ as follows: for $k \notin \{k(1), \dots, k(r)\}$ put $\lambda_{k,1} = \lambda_k$, while for $k \in \{k(1), \dots, k(r)\}$ put $\lambda_{k(2l-1),1} = \lambda_{k(2l-1)} + \alpha_l 2^{-N}$ and $\lambda_{k(2l),1} = \lambda_{k(2l)} - \alpha_l 2^{-N}$. Then $\lambda_{k,1}$ are dyadic fractions of length $N-1$ which satisfy the relations $\lambda_{k,1} \ge 0$ and $\sum_{k=m_i+1}^{m_{i+1}} \lambda_{k,1} = 1$. Let us use inequality (14) to estimate the quantity $\|\sum_{k=1}^{m_n} \lambda_k x_k - \sum_{k=1}^{m_n} \lambda_{k,1} x_k\|$:

$$\left\|\sum_{k=1}^{m_n} (\lambda_k - \lambda_{k,1}) x_k\right\| = \left\|\sum_{l=1}^{r/2} 2^{-N} \alpha_l (x_{k(2l-1)} - x_{k(2l)})\right\|$$

$$= 2^{-N}\left\|\sum_{l=1}^{r/2} \alpha_l g_l\right\| \le 2^{-N} C\left(\sum_{l=1}^{r/2} \|g_l\|^p\right)^{1/p}$$

$$= 2^{-N} C\left(\sum_{i=1}^{n} \sum_{m_{i-1} < k(2l) \le m_i} \|g_l\|^p\right)^{1/p} \le 2^{-N} C\left(\sum_{i=1}^{n} r_i d_i^p\right)^{1/p}$$

$$\leq 2^{-N} C \left(2^N \sum_{i=1}^{n} d_i^p \right)^{1/p} = C 2^{(\frac{1}{p}-1)N} \left(\sum_{i=1}^{n} d_i^p \right)^{1/p}.$$

Now apply to $\lambda_{k,1}$ the same procedure that we applied to the numbers λ_k. This yields dyadic fractions $\lambda_{k,2} \geq 0$ of length N_2, such that

$$\sum_{k=m_i+1}^{m_{i+1}} \lambda_{k,2} = 1, \quad i = 0, 1, \dots, n-1;$$

$$\left\| \sum_{k=1}^{m_n} \lambda_{k,1} x_k - \sum_{k=1}^{m_n} \lambda_{k,2} x_k \right\| \leq C 2^{(\frac{1}{p}-1)(N-1)} \left(\sum_{i=1}^{n} d_i^p \right)^{1/p}.$$

Iterating this procedure N times, we obtain dyadic fractions $\lambda_{k,N}$ of length zero, i.e., either zeros or ones. Because of the relation $\sum_{k=m_i+1}^{m_{i+1}} \lambda_{k,N} = 1$, we conclude that for $k \in \{m_i + 1, \dots, m_{i+1}\}$ exactly one of the numbers $\lambda_{k,N}$ is equal to one, while the remaining $\lambda_{k,N}$ are equal to zero. For each i denote by $s(i) \in \{m_i + 1, \dots, m_{i+1}\}$ the index for which $\lambda_{s(i),N} = 1$, and put $a_i = x_{s(i)}$. Then

$$\left\| \sum_{k=1}^{n} b_k - \sum_{k=1}^{n} a_k \right\| = \left\| \sum_{k=1}^{m_n} \lambda_k x_k - \sum_{k=1}^{m_n} \lambda_{k,N} x_k \right\| \leq \sum_{l=1}^{N} \left\| \sum_{k=1}^{m_n} \lambda_{k,l-1} x_k - \sum_{k=1}^{m_n} \lambda_{k,l} x_k \right\|$$

$$\leq C \sum_{l=1}^{N} 2^{(\frac{1}{p}-1)l} \left(\sum_{i=1}^{n} d_i^p \right)^{1/p} \leq C_1 \left(\sum_{i=1}^{n} d_i^p \right)^{1/p},$$

which completes the proof of the lemma. □

THEOREM 6. *If X has infratype $p > 1$, then $X \in$* **COIN**.

PROOF. Let C_1 be the constant provided for X by Lemma 3. Further, let $f\colon [0,1] \to X$ be an arbitrary bounded function, $M(f) = M$, and $x \in \mathrm{WI}(f)$. Pick a sequence (Γ_n, T_n) with $d(\Gamma_n) \to 0$, such that $S_n = S(f, \Gamma_n, T_n) \to x$ weakly. Take an arbitrary $\varepsilon > 0$ and choose $m \in \mathbf{N}$ such that $4Mm^{\frac{1}{p}-1}C_1 < \varepsilon$. Divide $[0,1]$ into m equal segments and denote by $S_{k,n}$ the part of the integral sum S_n which corresponds to the segments Δ_i that lie completely inside the interval $[(k-1)/m, k/m]$. As we already did several times before, without affecting the weak convergence $S_n \to x$ we may assume that all the points k/m, $k = 1, \dots, m$ are partition points for Γ_n. Then $S_n = \sum_{k=1}^{m} S_{k,n}$. Now we use a theorem of Mazur which asserts that for any sequence $\{y_n\}$ that converges weakly to x there exists a sequence of convex combinations of the elements y_k that converges to x in norm. Pick coefficients $\lambda_i \geq 0$ such that $\sum_{i=1}^{N} \lambda_i = 1$ and $\| \sum_{i=1}^{N} \lambda_i S_i - x \| \leq \varepsilon/2$. Set $b_k = \sum_{k=1}^{N} \lambda_n S_{k,n}$ and $A_k = \{S_{k,n}\}_{n=1}^{N}$. Clearly, $b_k \in \operatorname{conv} A_k$ and the diameter of the set A_k does not exceed $2M/n$. Using Lemma 3, choose numbers $n(k)$ such that

$$\left\| \sum_{k=1}^{m} S_{k,n(k)} - \sum_{k=1}^{n} b_k \right\| \leq C_1 2M m^{\frac{1}{p}-1} < \frac{\varepsilon}{2}.$$

Since $S = \sum_{k=1}^{m} S_{k,n(k)}$ is an integral sum (composed of "pieces" of the sums S_n) and since the element $\sum_{k=1}^{n} b_k = \sum_{i=1}^{N} \lambda_i S_i$ is close to x, the last inequality means that $\|x - S\| < \varepsilon$, i.e., x can be arbitrarily well approximated in norm by integral sums of the function f. Therefore, $x \in \mathrm{I}(f)$. □

Thus, we have reached the conclusion that classes of spaces that in some sense are opposite (the B-convex spaces and the spaces with the Schur property) belong to **COIN**. Before we give new examples of spaces with the coincidence property, let us show that there are spaces that do not have this property.

EXAMPLE 3. Consider again the sets $E_n \subset [0,1]$ constructed in Example 2. Let $t \mapsto e_t$ be the bijective mapping of the set $E = \bigcup_{n=1}^{\infty} E_n$ onto the set of the unit vectors of the standard basis of the space l_1, and let $\{g_n\}_{n=1}^{\infty}$ be an othonormal basis in l_2. Define the function $f\colon [0,1] \to l_1 \oplus l_2$ as follows:

$$f(t) = g_{n+1} + \frac{1}{2^n} \sum_{\alpha \in E_n} e_\alpha - e_t$$

if $t \in E_n$, and $f(t) = g_1$ if $t \notin E$. Obviously, $M(f) \leq 3$. If one takes Γ_n to be the partition of $[0,1]$ into 2^n equal segments, then each of these segments will contain a point from E_n. Consequently, $S(f, \Gamma_n, E_n) = \frac{1}{2^n} \sum_{t \in E_n} f(t) = g_{n+1}$. Since $g_n \to 0$ weakly, it follows that $0 \notin \mathrm{WI}(f)$. On the other hand, arguing much in the same way as in the proof of Theorem 2 we conclude that $0 \notin \mathrm{I}(f)$. Therefore, $l_1 \oplus l_2 \notin$ **COIN**.

CONJECTURE 1. *If X contains an isomorphic copy of l_1 but does not have the Schur property, then $X \notin$* **COIN**.

Combining theorems 5 and 6 we recover the already familiar Theorem 1: the class **CONV** contains all B-convex spaces. To ehxibit new representatives of **CONV** we shall introduce a new property of Banach spaces, called weak B-convexity, then study it and prove that the reflexive weakly B-convex spaces belong to **COIN**.

DEFINITION 6. A Banach space X is said to be *weakly B-convex* if there exist numbers $n \in \mathbf{N}$ and $\varepsilon > 0$ such that, for any finite collection $\{A_i\}_{i=1}^{n}$ of weakly convergent to zero sequences of elements of the unit ball of X one can find elements $\{x_i\}_{i=1}^{n}$, $x_i \in A_i$, $i = 1, \ldots, n$, for which $\|\sum_{i=1}^{n} x_i\| \leq n(1-\varepsilon)$.

In what follows, if $A \subset X$ is a bounded set we will denote $\sup\{\|x\| : x \in A\}$ simply by $\|A\|$.

DEFINITION 7. A Banach space X is said to have *weak type p with constant C* if for any finite collection $\{A_i\}_{i=1}^n$ of weakly convergent to zero sequences of elements of X one can find elements $\{x_i\}_{i=1}^n$, $x_i \in A_i$, $i = 1, \dots, n$, for which

$$\left\| \sum_{i=1}^{n} x_i \right\| \leq C \left(\sum_{i=1}^{n} \|A_i\|^p \right)^{1/p}.$$

Obviously, a space that has weak type $p > 1$ is weakly B-convex. Following the scheme used by G. Pisier in [12] to prove that every B-convex spaces has an infratype $p > 1$ one can obtain the following result:

THEOREM 7. *Every weakly B-convex space has a weak type $p > 1$.*

To justify the term "weak B-convexity" let us show that this notion is indeed weaker than that of B-convexity.

THEOREM 8. *If X has infratype $p > 1$, then X has weak type p.*

PROOF. Let C_1 be the constant appearing in Lemma 3, and let $\{A_i\}_{i=1}^n$ be an arbitrary finite collection of weakly convergent to zero sequences of elements of X. By Mazur's theorem on weakly convergent sequences, for any $\varepsilon > 0$ there exist elements $b_i \in \operatorname{conv} A_i$, $i = 1, \dots, n$, such that $\|b_i\| < \varepsilon$ for all i. Lemma 3 yields elements $x_i \in A_i$ such that

$$\left\| \sum_{i=1}^{n} x_i - \sum_{i=1}^{n} b_i \right\| \leq C_1 \left(\sum_{i=1}^{n} (2\|A_i\|)^p \right)^{1/p}.$$

Then

$$\left\| \sum_{i=1}^{n} x_i \right\| \leq n\varepsilon + 2C_1 \left(\sum_{i=1}^{n} \|A_i\|^p \right)^{1/p}.$$

Since ε is arbitrary, this means that X has weak type p with constant $3C_1$. □

The class of weakly B-convex spaces is considerably wider than that of B-convex spaces: for instance, it contains all the spaces with the Schur property as well as the space c_0. Let us give an example showing that among the reflexive spaces there are many weakly B-convex spaces that are not B-convex.

EXAMPLE 4. Let $X_1, \dots, X_n, \dots$ be a sequence of finite-dimensional normed spaces. Consider the space $Y = (\sum X_n)_2$, i.e., the space of sequences of the form $a = (a_1, \dots, a_n, \dots)$, where $a_i \in X_i$, $\sum_{i=1}^{\infty} \|a_i\|^2 < \infty$, equipped with the norm $\|a\| = \left(\sum_{i=1}^{\infty} \|a_i\|^2\right)^{1/2}$. Y is a reflexive space. At the same time, a space of this type is not necessarily B-convex (for example, take $X_n = l_1^{(n)}$). Let us show that Y is necessarily weakly B-convex with $n = 2$ and any $\varepsilon < 1 - \sqrt{2}/2$.

Choose a $\delta > 0$ such that $\sqrt{2}+\sqrt{2\delta} < 2(1-\varepsilon)$. Let A and B be two sequences of elements of Y that converge weakly to zero, $\|A\| = \|B\| = 1$, and let $a = (a_i)_{i=1}^{\infty}$ be an arbitrary element of A. Fix a number N such that $\sum_{i=N+1}^{\infty} \|a_i\|^2 < \delta$. Since the sequence B is weakly convergent, one can choose an element $b \in B$, $b = (b_i)_{i=1}^{\infty}$, such that $\sum_{i=1}^{N} \|b_i\| < \delta$ (weak convergence implies coordinate-wise convergence). Then

$$\|a+b\| = \left(\sum_{i=1}^{\infty} \|a_i + b_i\|^2\right)^{1/2}$$

$$\leq \left(\sum_{i=1}^{N} \|a_i\|^2 + \sum_{i=N+1}^{\infty} \|b_i\|^2\right)^{1/2} + \left(\sum_{i=N+1}^{\infty} \|a_i\|^2 + \sum_{i=1}^{N} \|b_i\|^2\right)^{1/2}$$

$$\leq \sqrt{2} + \sqrt{2\delta} < 2(1-\varepsilon).$$

This means that the space Y is weakly B-convex. A more careful examination shows that Y has weak type $p = 2$.

THEOREM 9. *Let X have weak type $p > 1$ with constant C. It X is reflexive, then $X \in$* **COIN**.

PROOF. Let $f\colon [0,1] \to X$ be a bounded function, $M(f) = M$, $x \in \mathrm{WI}(f)$, and let $S_n = S(f, \Gamma_n, T_n)$, $d(\Gamma_n) \to 0$, be a sequence of integral sums of f that converges weakly to x. As in the proof of Theorem 6, let us divide $[0,1]$ into m equal segments and denote by $S_{k,n}$ the part of the integral sum that corresponds to the segments $\Delta_i \subset [(k-1)/m, k/m]$. Using the reflexivity of X we can choose a sequence of indices n_j such that the weak limit $\mathrm{w} - \lim_{j\to\infty} S_{k,n_j}$ will exist for all $k = 1, \dots, m$. Denote $\mathrm{w} - \lim_{j\to\infty} S_{k,n_j}$ by x_k. Clearly, $\sum_{k=1}^{m} x_k = x$. Further, since

$$\|S_{k,n_j} - x_k\| \leq 2M/m, \quad \mathrm{w} - \lim_{j\to\infty} (S_{k,n_j} - x_k) = 0,$$

we can choose numbers $\overline{n}_k$ such that

$$\left\|\sum_{k=1}^{m} (S_{k,\overline{n}_k} - x_k)\right\| \leq C2\,\frac{M}{m}\,m^{1/p}.$$

As in the proof of Theorem 6, the quantity $\sum_{k=1}^{m} S_{k,\overline{n}_k}$ can be regarded as some integral sum $\overline{S}_m$ of the function f. Then the last inequality reads $\|\overline{S}_m - x\| \leq 2CMm^{\frac{1}{p}-1}$ and since $p > 1$, we conclude that $\lim_{m\to\infty} \|\overline{S}_m - x\| = 0$. This means that $x \in \mathrm{I}(f)$, which completes the proof of the theorem. □

Combining theorems 5, 7, and 9 we see deduce that every reflexive weakly B-convex space has the **CONV** property. Notice also that the property **CONV** is inherited by subspaces as well as by quotient spaces.

CONJECTURE 2. *The properties* **CONV** *and* **WCONV** *are equivalent.*

QUESTION. Does the space c_0 have the property **CONV** or **WCONV**?

Exercises

1. Show that if X is infinite-dimensional, then from the fact that $\mathrm{I}(f)$ consists of a single point it does not necessarily follow that f is Riemann-integrable.

2. Show that if f take values in some compact set $K \subset X$, then $\mathrm{I}(f)$ is a convex set.

3. Show that any convex, closed, bounded subset of a Banach space with the power of the continuum can serve as the set $\mathrm{I}(f)$ for some function f.

4. Give an example of a space that is not weakly B-convex.

5. Let $X \in$ **WCONV** be a reflexive space. Show that any quotient space of X also belongs to **WCONV**.

6. Using the **COIN** property, provide an example of a nonseparable space X such that for any function $f\colon [0,1] \to X$ with $M(f) < \infty$ the set $\mathrm{I}(f)$ is nonempty.

Supplementary List of References

[1] H. W. Ellis, *On the limits of Riemann sums*, J. London Math. Soc. **34** (1959), no. 1, 93–100 [MR **20** #7084].

[2] I. Halperin and N. Miller, *An inequality of Steinitz and the limits of Riemann sums*, Trans. Roy. Soc. Canada Sect. III. (3) **48** (1954), 27–29 [MR **16**, 596a].

[3] P. Hartman, *On the limits of Riemann approximating sums*, Quart. J. Math. Oxford Ser. **18** (1947), 124–127 [MR **9**, 137f].

[4] R. L. Jeffery, *Limit points of Riemann sums*, Trans. Roy. Soc. Canada Sect. III. (3) **44** (1950), 43–49 [MR **12**, 487d].

[5] V. M. Kadets, *Riemann integral sums and the geometry of a Banach space*, Dokl. Akad. Nauk Ukrain. SSR Ser. A (1987), no. 12, 12–14 (Russian) [MR 89d:46018].

[6] V. M. Kadets, *The domain of weak limits of Riemann integral sums of an abstract function*, Izv. Vyssh. Uchebn. Zaved. Mat. (1988), no. 9, 39–46 (Russian) [MR 90k:46099].

[7] V. M. Kadets, *The starness of the domain of limits of Riemann integral sums of a vector-valued function*, in: Operator Theory, Subharmonic Functions, "Naukova Dumka", Kiev, 1991, pp. 60–67 (Russian) [MR 93m:26046].

[8] V. M. Kadets and M. I. Kadets, *Conditions for the convexity of the set of limits of Riemann sums of a vector-valued function*, Mat. Zametki **35** (1984), no. 2, 161–167 (Russian) [MR 85k:46010].

[9] D. P. Mil'man and V. D. Mil'man, *Some properties of non-reflexive Banach spaces*, Mat. Sb. **65(107)** (1964), no. 4, 486–497 (Russian) [MR **30** #1383].

[10] M. Nakamura and I. Amemiya, *On the limits of Riemann sums of functions in Banach spaces*, J. Fac. Sci. Hokkaido Univ. Ser. I **19** (1966), no. 3,4, 135–145 [MR **34** #6023].

[11] M. I. Ostrovskii, *Domains of sums of conditionally convergent series in Banach spaces*, Teor. Funktsii Funktsional. Anal. i Prilozhen. **46** (1986), no. 2, 77–85 (Russian) [MR 88i:46024].

[12] G. Pisier, *Sur les espaces de Banach qui ne contiennent pas uniformément de* $l_n^{(1)}$, C. R. Acad. Sci. Paris, Ser. A-B **277** (1973), A991–A994 [MR **48** #11998].

[13] H. P. Rosenthal, *Some recent discoveries in the isomorphic theory of Banach spaces*, Bull. Amer. Math. Soc. **84** (1978), no. 5, 803–831 [MR 80d:46023].

COMMENTS ON THE EXERCISES

Chapter 1

EX. 1.1.1. False; for more details, see §1 of Chapter 2.

EX. 1.1.2. Let $A \subset \mathbf{N}$. Denote by $\rho(A)$ the smallest number s of segments of the form $[n_1, m_1] = \{n_1, n_1+1, n_1+2, \dots, m_1-1, m_1\}$ such that A can be written as $\bigcup_{i=1}^{s}[n_i, m_i]$. For a permutation $\pi\colon \mathbf{N} \to \mathbf{N}$ denote $\rho(\pi) = \sup\{\rho(\pi^{-1}[n,m])\colon n, m \in \mathbf{N},\ n \le m\}$.

THEOREM (A. S. KRONROD, 1946). *The permutation π is not convergence-modifying if and only if $\rho(\pi) < \infty$.*

Let us sketch the proof proposed in 1985 by V. M. Ogranovich, one of the participants in the special course read that year by M. I. Kadets at Khar'kov University.

Let E be the space of numerical sequences $e = \{\nu_1, \nu_2, \dots\}$ for which the series $\sum_{i=1}^{\infty} \nu_i$ converges. Equip E with the norm

$$\|e\| = \sup\left\{\left|\sum_{i=n}^{m} \nu_i\right| : n, m \in \mathbf{N},\ n \le m\right\}.$$

Then E is a Banach space (which, incidentally, is isomorphic to c_0). Let D_π be the set of all elements $e \in E$ for which the series $\sum_{i=1}^{\infty} \nu_{\pi(i)}$ converges. Define an operator $T_\pi\colon D_\pi \to E$ by the rule $T_\pi((\nu_1, \nu_2, \dots)) = (\nu_{\pi(1)}, \nu_{\pi(2)}, \dots)$. Then T_π is closed; its domain D_π contains all the finite sequences and hence is dense in E. Thus, for D_π to coincide with the whole space E (i.e., for π to be not convergence-modifying) it is necessary and sufficient that T_π be a bounded operator. It remains to show that $\|T_\pi\| = \rho(\pi)$. Suppose that $n < m$ and the set $\pi^{-1}[n, m]$ is represented in the form

$$\pi^{-1}[n,m] = \bigcup_{i=1}^{\rho(\pi^{-1}[n,m])} [n_i, m_i], \quad n_1 \le m_1 < n_2 \le m_2 < \dots.$$

Consider the element $e \in E$ for which $\nu_{m_i} = 1$, $\nu_{m_i+1} = -1$, $i = 1, \dots, \rho(\pi^{-1}[n,m])$, and the remaining coordinates are equal to 0. Then $\|e\| = 1$ and

$$\|T_\pi e\| \ge \left|\sum_{i=n}^{m} \nu_{\pi(i)}\right| = \sum_{i=1}^{\rho(\pi^{-1}[n,m])} \nu_{m_i} = \rho(\pi^{-1}[n,m]).$$

Since the segment $[n,m]$ is arbitrary, we conclude that $\|T_\pi\| \geq \rho(\pi)$. The opposite inequality $\|T_\pi\| \leq \rho(\pi)$ is an immediate consequence of the definition of the norm in E.

EX. 1.1.3. A permutation is sum-modifying if $\lim_{n\to\infty} \rho(\pi^{-1}[1,n]) = \infty$ (in the notations of the preceding comment). We cannot offer a "nice" proof.

EX. 1.1.4. Combine the answers to the two preceding exercises. One can also give a direct proof (M. I. Kadets):

Let E and T_π be as in the comment to Exercise 1.1.2. Define a functional $f \in E^*$ by $f(e) = \sum_{i=1}^\infty \nu_i$. If the permutation is not convergence-modifying, then $f \circ T_\pi$ is a bounded functional. Since $f \circ T_\pi = f$ for all elements with finitely many nonzero coordinates, and since the subset of these elements is dense in E, $f(T_\pi(e)) = f(e)$ for all $e \in E$, i.e., $\sum_{i=1}^\infty \nu_{\pi(i)} = \sum_{i=1}^\infty \nu_i$ for any convergent series $\sum_{i=1}^\infty \nu_i$.

EX. 1.1.5. Let $\sum_{k=1}^\infty x_k = 0$. Denote $u_n = \sum_{k=1}^n x_{2k}$ and $v_n = \sum_{k=1}^n x_{2k-1}$. Then $\lim_{n\to\infty}(u_n + v_n) = 0$. For the permutation π we have $\sum_{k=1}^{3n} x_{\pi(k)} = u_n + v_{2n}$. Hence, $\lim_{n\to\infty}(u_n + v_{2n}) = x$. For the permutation σ we have

$$\sum_{k=1}^{5n} x_{\sigma(k)} = u_n + v_{4n} = (u_n + v_{2n}) + (u_{2n} + v_{4n}) - (u_{2n} + v_{2n}).$$

Since $\lim_{n\to\infty}(u_n + v_{2n}) = x$, $\lim_{n\to\infty}(u_{2n} + v_{4n}) = x$, and $\lim_{n\to\infty}(u_{2n} + v_{2n}) = 0$, it follows that $\lim_{n\to\infty} \sum_{k=1}^{5n} x_{\sigma(k)} = 2x$. Since the general term of the series tends to 0, we conclude that the series $\sum_{k=1}^\infty x_{\sigma(k)}$ converges and its sum is equal to $2x$ (V. M. Kadets).

EX. 1.3.2. Show that the closure of the linear span in $L_1[0,1]$ of the sequence $r_n(t) = \operatorname{sign} \sin(2^n \pi t)$ is isomorphic to l_2 (use Khinchin's inequality from §4 of Chapter 3). Using this fact, transfer Example 1.3.1 to $L_1[0,1]$. In the space $L_p[0,1]$ with $1 \leq p < \infty$ one can argue in the same manner; $L_\infty[0,1]$ and $C[0,1]$ also contain isomorphic copies of l_2. The most difficult case is that of the space l_1.

Chapter 2

EX. 2.1.1. Modify the method of choosing the coefficients θ_i for which $\theta_i \neq \bar{t}_i$.

EX. 2.1.2. By assumption, there are permutations π, σ, and sequence of numbers $n(1) < n(2) < n(3) < \dots$, $m(1) < m(2) < m(3) < \dots$, such that

$$\lim_{k\to\infty} \left\| \sum_{k=1}^{n(k)} x_{\pi(i)} \right\| = 0, \quad \lim_{k\to\infty} \left\| x - \sum_{k=1}^{m(k)} x_{\sigma(i)} \right\| = 0.$$

Using this fact, choose sets of indices $M_1 \subset N_1 \subset M_2 \subset N_2 \subset \dots$ for which

$$\left\| \sum_{i \in M_k} x_i - x \right\| \leq \frac{1}{2^k}, \quad \left\| \sum_{i \in N_k} x_i \right\| \leq \frac{1}{2^k}, \quad \bigcup_{i=1}^{\infty} M_i = \mathbf{N}.$$

Consider the sets $R_1 = M_1 \cup (M_2 \setminus N_1)$, $R_2 = M_3 \cup (M_4 \setminus N_3)$, $R_3 = M_5 \cup (M_6 \setminus N_5)$, $\dots$, $R_n = M_{2n-1} \cup (M_{2n} \setminus N_{2n-1})$, $\dots$. Clearly, $R_1 \subset R_2 \subset R_3 \subset \dots$, $\bigcup_{i=1}^{\infty} R_i = \mathbf{N}$, and

$$\left\| 2x - \sum_{i \in R_n} x_i \right\| \leq \left\| x - \sum_{i \in M_{2n-1}} x_i \right\| + \left\| \sum_{i \in N_{2n-1}} x_i \right\| + \left\| x - \sum_{i \in M_{2n}} x_i \right\|$$

$$\leq \frac{1}{2^{2n-1}} + \frac{1}{2^{2n-1}} + \frac{1}{2^{2n}}. \tag{1}$$

Define a new permutation ν as follows: write first the elements of R_1, then those of $R_2 \setminus R_1$, $R_3 \setminus R_2$, and so on. Denote by l_k the number of elements in R_k. Then $\sum_{i=1}^{l_k} x_{\nu(i)} = \sum_{i \in R_k} x_i$, $k = 1, 2, 3, \dots$ and (1) implies that

$$\lim_{k \to \infty} \left\| 2x - \sum_{i=1}^{l_k} x_{\nu(i)} \right\| = 0.$$

EX. 2.1.3. See, e.g., [30]. Yet another proof of Steinitz's theorem in this simplified formulation is based on P. Lévy's idea [50] of considering the set of "limit directions of a series." A correct exposition can be found in Steinitz's classical work or in D. O. Shklyarskii's paper [85].

EX. 2.1.4. The characterization is that the following two conditions are simultaneously satisfied: a) $\lim_{k \to \infty} x_k = 0$; b) for any functional f both series $\sum_{k=1}^{\infty} [f(x_k)]^+$ and $\sum_{k=1}^{\infty} [f(x_k)]^-$ either converge, or diverge. In the infinite-dimensional case the situation is quite different (see [42]).

EX. 2.2.2. In Exercise 2.2.1 each of assertions 1) and 2) is equivalent to the space being finite-dimensional, independently of the completeness assumption. In contrast to Exercise 2.2.1, the present problem is quite delicate. We only know a proof based on Dvoretzky's Theorem 6.2.1.

EX. 2.3.1. Let $\{e_k\}_{k=1}^{\infty}$ be an orthonormal sequence in a Hilbert space. Then the needed example is provided by the series

$$e_1 - e_1 + \frac{1}{\sqrt{2}} e_2 - \frac{1}{\sqrt{2}} e_2 + \frac{1}{\sqrt{3}} e_3 - \frac{1}{\sqrt{3}} e_3 + \dots.$$

If here one selects a large group of terms with the sign $+$, then one with the sign $-$, then another large group with the sign $+$, then one with the sign $-$, and so on, one obtains a perfectly divergent series.

Chapter 3

Ex. 3.1.1. See Theorem 6.4.1. There also exists a proof that does not resort to the theorem on basic sequences.

Ex. 3.1.2. The series converges absolutely at all points $x < 1$, but converges conditionally at $x = 1$.

Ex. 3.1.4. This result was obtained by N. N. Bogdan in her MSc Thesis (Zaporozhye University, Ukraine, 1992, not published).

Ex. 3.1.5. On the unit sphere of the space Y pick a countable dense set of elements $\{y_i\}_{i=1}^{\infty}$. Let $\sum_{i=1}^{\infty} \varepsilon_i$ be an arbitrary conditionally convergent numerical series. The set of terms of the sought-for series is the union of the following sets:

$$\{\varepsilon_k y_1\}_{k=1}^{\infty}, \ \{\frac{1}{2}\varepsilon_k y_2\}_{k=1}^{\infty}, \ \dots \ \{\frac{1}{2^n}\varepsilon_k y_n\}_{k=1}^{\infty}, \ \dots .$$

The reader will encounter no difficulty in proving that for this series it holds that $\mathrm{SR}(\sum_{k=1}^{\infty} x_k) = Y$.

Ex. 3.1.6. A weakly convergent sequence in L_2 may diverge at every point: the Rademacher functions r_n take the values ± 1, but the weak limit is identically equal to zero. To solve the exercise one should use precisely such sequences of partial sums that "converge in the mean."

Ex. 3.1.8. Such examples were constructed for the first time in P. A. Kornilov's PhD dissertation. A proof that works for arbitrary spaces is given in Chapter 7.

Ex. 3.1.9. See [90].

Ex. 3.2.1. See P. A. Kornilov's paper [48] or V. M. Kadets' paper [40]. The results were obtained at about the same time and independently, but Kornilov's work was published earlier.

Ex. 3.2.2. LPR has a "group structure" (H. Hadwiger [97]): if $x, y, z \in$ LPR, then $x + y - z \in$ LPR. In other words, LPR is a shifted group with respect to addition. The proof is similar to the comment on Exercise 2.1.2.

If one considers only series whose general term tends to zero, LPR has another interesting property (V. M. Kadets): if LPR contains a subspace H_1 of finite codimension, then LPR itself is a subspace.

Proof. In view of the first part of this comment, it suffices to show that for any $a \in \mathrm{LPR} \setminus H_1$ the entire line λa, $\lambda \in \mathbf{R}$ is contained in LPR. Consider the canonical projection $T\colon X \to X/H_1$. If $\sum_{i=1}^{\infty} x_i$ is the original series, then $\sum_{i=1}^{\infty} T(x_i)$ is already a series in a finite-dimensional space. From Steinitz's lemma it readily follows that $\mathrm{SR}(\sum_{i=1}^{\infty} T(x_i)) = \mathrm{LPR}(\sum_{i=1}^{\infty} T(x_i))$. Consequently, the series $\sum_{i=1}^{\infty} T(x_i)$ has a rearrangement that converges to 0, as well as one that converges to $T(a)$, and, by Steinitz's theorem, there exist a permutation π of the terms for which the series $\sum_{i=1}^{\infty} T(x_{\pi(i)})$ converges to $T(\lambda a)$. Hence, since $H_1 \subset \text{LPR}(\sum_{i=1}^{\infty} x_i)$, the element λa can be arbitrarily well approximated by partial sums of the form $\sum_{i \in A} x_i$; moreover, the index set A can be selected to contain any prescribed finite set of indices. This means that $\{\lambda a \colon \lambda \in \mathbf{R}\} \subset \text{LPR}$.

We cannot provide a precise characterization of the sets that can serve as LPR for a series whose general term tends to zero.

Chapter 4

EX. 4.1.1. Proceed by analogy with the Dvoretzky-Rogers theorem, but instead of the inscribed ellipsoid of maximal volume consider the circumscribed ellipsoid of minimal volume.

EX. 4.2.3. Consider series of functions with disjoint supports.

EX. 4.2.4. Yet another proof of this fact can be found in §2 of Chapter 8.

EX. 4.2.5. Consider the operator T that maps the unit vectors of the standard basis of l_1 into the functions r_n: $Te_n = r_n$. Show that T is an isometry.

EXS. 4.3.8 AND 4.3.9. For a detailed exposition of the theory of nuclear spaces consult the monograph [69].

Chapter 5

EX. 5.1.2. Take $X = l_2$ and $Y = c_0$.

EX. 5.1.3. Show that the parallelogram equality holds in X.

EX. 5.1.4. Let $\mathcal{S}$ be the set of all two-dimensional normed spaces, equipped with the metric $\ln d(X, Y)$. Let $\{A_i\}_{i=1}^{\infty}$ and $\{B_i\}_{i=1}^{\infty}$ be two dense sequences in $\mathcal{S}$ such that the unit sphere of A_i [resp. B_i] is a polyhedron [resp. a smooth curve] for all $i \geq 1$. Then $X = (\sum_{i=1}^{\infty} A_i)_2$ and $Y = (\sum_{i=1}^{\infty} B_i)_2$ provide the needed spaces.

EX. 5.1.5. See [52].

EX. 5.1.6. Represent the unit ball of the space X as the intersection of a finite number of half-spaces. Correspondingly, the norm in X can be written as $\|x\| = \max_{1 \leq i \leq m} |f_i(x)|$, where $f_i \in X^*$ for $i = 1, \ldots, m$. Then the operator $T \colon X \to l_\infty^{(m)}$ acting as $Tx = (f_i(x))_{i=1}^m$ is an isometric embedding.

EX. 5.1.7. Show that if the unit ball of X is a polygon, then the norm can be written as $\|x\| = \sum_{i=1}^m |f_i(x)|$, where $f_i \in X^*$ for $i = 1, \ldots, m$. Now argue by analogy with the preceding exercise.

EX. 5.1.8. Show that the M-cotype is preserved.

EX. 5.1.9. $d\left(l_2^{(n)}, l_p^{(n)}\right) = n^{\left|\frac{1}{2} - \frac{1}{p}\right|}$ (see [11], [28]).

EX. 5.1.10. These metric compact spaces $\mathcal{M}_n$ are known as the *Minkowski compacta*. An interesting question is what is the diameter d_n of $\mathcal{M}_n$? E. D. Gluskin has shown that $\log n - d_n \to 0$ as $n \to \infty$.

EX. 5.1.11. Choose a countable set $\mathcal{A}$ of finite-dimensional subspaces of X with the property that, for any finite-dimensional subspace V of X there are elements of $\mathcal{A}$ arbitrarily close to V in the Banach-Mazur distance. Now take $Y = \overline{\mathrm{Lin}} \bigcup_{A\in\mathcal{A}} A$.

EX. 5.1.12. See [86].

EX. 5.2.3. Take $X = \left(\sum_{n=1}^{\infty} l_{2+\frac{1}{n}}^{(k_n)}\right)_2$, where the numbers k_n are large enough (say, $k_n = 2^n$).

EX. 5.3.2. If the space X is not C-convex, then $c_0 \xrightarrow{\mathrm{f}} X$. By Corollary 5.1.1, $l_1 \xrightarrow{\mathrm{f}} c_0$. We conclude that $l_1 \xrightarrow{\mathrm{f}} X$, and hence X is not B-convex.

EX. 5.3.3. One can actually prove the following more general fact: any separable Banach space is a quotient space of l_1. Here is a sketch of a proof:

Let X be separable. Choose a dense sequence $\{x_i\}_{i=1}^{\infty}$ in $S(X)$. Define an operator $T\colon l_1 \to X$ by the rule $Te_i = x_i$, where $\{e_i\}_{i=1}^{\infty}$ is the standard basis of l_1. Then $\|T\| = 1$ and T is a surjection. Consequently, $l_1/\mathrm{Ker}\, T = X$.

EX. 5.3.4. See [73]. If you find a relatively elementary proof, then you should publish it.

EX. 5.3.5. This is analogous to the corresponding result on the M-cotype.

EX. 5.3.6. Show that the spaces X and $l_2(X)$ have the same type.

EX. 5.3.7. X and $l_2(X)$ have the same cotype.

EX. 5.3.8. See [13, Chapter 3].

EX. 5.3.9. This is a theorem of M. I. Kadets [32]; see [13, Chapter 3].

EX. 5.3.10. See [13].

Chapter 6

EX. 6.1.2. Use the fact that for a convex function f the set $\{(x,t)\colon x \in X,\ t \in \mathbf{R}\}$ is convex. Apply the Hahn-Banach theorem.

EXS. 6.1.3–6.1.7. See [13].

EX. 6.2.1. If X is an infinite-dimensional subspace of c_0, then X contains a sequence of elements from $S(c_0)$ that is arbitrarily close to being disjoint. Therefore X contains a subspace isomorphic to c_0, and hence X is not reflexive. An analogous assertion (every infinite-dimensional subspace contains another subspace that is isomorphic to the whole space) holds true for the space l_p with $p < \infty$, $p \neq 2$.

Ex. 6.2.2. See [99].

Ex. 6.3.1. In c_0 and l_p the standard basis is a basis; in $C[0,1]$ the Schauder system is a basis; and in $L_p[0,1]$ the Haar system is a basis. For more details see [55, Chapter 3, §6]; the most fundamental exposition is found in [83].

Ex. 6.4.1. Sufficiency is an obvious corollary of Riemann's theorem. Necessity follows from the Banach-Steinhaus theorem.

Ex. 6.4.2. No. If $\{e_n\}$ is the standard basis in c_0, then a counterexample is provided by the series

$$e_1 + \frac{1}{2}e_2 + \frac{1}{2}e_2 + \frac{1}{3}e_3 + \frac{1}{3}e_3 + \frac{1}{3}e_3 + \frac{1}{4}e_4 + \dots .$$

Ex. 6.4.3. One needs to show that every weakly unconditionally convergent series in a Banach space converges unconditionally in norm. The proof can be carried out by reductio ad absurdum. Suppose $\sum_{i=1}^{\infty} x_i$ is weakly unconditionally convergent but diverges in the strong topology. By analogy with the proof of the implication (a) $\Rightarrow$ (d) in Theorem 6.4.3, one obtains series $\sum_{i=1}^{\infty} y_n$ and $\sum_{i=1}^{\infty} z_n$ which, in addition to the properties from the proof of that implication, are also weakly convergent. Since the sequence $\{z_i\}$ is equivalent to the standard basis of c_0, we reach a contradiction.

Chapter 7

Ex. 7.1.1. Take $\lambda_i = \frac{1}{2}$. Then $x - \sum_{i=1}^{n} \theta_i x_i = \sum_{i=1}^{n} \pm\frac{1}{2}x_i$.

Ex. 7.1.2. The following two assertions are equivalent:

(1) For any perfectly divergent series $\sum_{i=1}^{\infty} x_i$ in X the series $\sum_{i=1}^{\infty} \|x_i\|^p$ diverges.

(2) There exists a $C > 0$ such that, for any $n \in \mathbf{N}$ and any collection $\{x_i\}_{i=1}^{n} \subset X$,

$$\min_{\nu_i=\pm 1} \max_{1\le k\le n} \left\| \sum_{i=1}^{k} \nu_i x_i \right\| \le C \left(\sum_{i=1}^{n} \|x_i\|^p \right)^{1/p} .$$

Ex. 7.2.2. Define an operator $T\colon X \to X$ by the rule $Te_i = y_i$. Then T is bounded; moreover, $\|T - I\| < 1$ (here I is the identity operator). Hence, T is an isomorphism and since the image of a basis under an isomorphism is a basis, we conclude that $\{y_i\}_{i=1}^{\infty}$ is a basis.

Ex. 7.2.3. First prove that this is true for Hilbert spaces, and then use Theorem 7.2.2.

Ex. 7.3.3. Use the fact that in this case $l_p \overset{\mathrm{f}}{\Rightarrow} X$, and hence $L_p \overset{\mathrm{f}}{\Rightarrow} X$ (see the list of results on type and cotype)

Chapter 8

Ex. 8.1.1. Use Theorem 7.2.2 and the fact that a strongly continuous operator is weakly continuous. Then the example of a series in l_2 for which $\{0, u\} \subset \mathrm{SR}$, but $\frac{1}{2} \notin \mathrm{WSR}$ can be carried over to any infinite-dimensional Banach space.

Ex. 8.2.1. One can construct an example similar to Nikishin's example.

Ex. 8.3.1. See §1 of Chapter 7 (property (A_1) implies the existence of an infratype) and Lemma 2.3.3.

Ex. 8.3.3. Show that $\mathbf{R}^{[0,1]}$ contains a subspace isomorphic to l_2, equipped with the weak topology.

Ex. 8.3.4. A space with the strongest locally-convex topology.

REFERENCES

[1] W. Banaszczyk, *The Steinitz constant of the plane*, J. Reine Angew. Math. **373** (1987), 218–220 [MR 88e:52016].

[2] W. Banaszczyk, *The Steinitz theorem on rearrangement of series for nuclear spaces*, J. Reine Angew. Math. **403** (1990), 187–200 [MR 90k:46010].

[3] W. Banaszczyk, *Additive Subgroups of Topological Vector Spaces*, Lecture Notes in Math., Vol. 1466, Springer-Verlag, Berlin, 1991 [MR 93b:46005].

[4] I. Bàràny, *A vector-sum theorem and its applications to improving flow shop guarantees*, Math. Oper. Res. **6** (1981), no. 3, 445–452 [MR 82i:90057].

[5] I. Bàràny and V. S. Grinberg, *On some combinatorial questions in finite-dimensional spaces*, Linear Algebra Appl. **41** (1981), 1–9 [MR 84f:05001].

[6] B. Beauzamy, *Introduction to Banch Spaces and their Geometry*, 2nd edition, North-Holland, Amsterdam-New York, 1985 [MR 88f:46021].

[7] A. Beck, *A convexity condition in Banach spaces and the strong law of large numbers*, Proc. Amer. Math. Soc. **13** (1962), 329–334 [MR **24** #A3681].

[8] I. S. Belov and Ya. N. Stolin, *An algorithm for the single-route scheduling problem*, in: Mathematical Economics and Functional Analysis, "Nauka", Moscow, 1974, pp. 248–257 (Russian); [MR **56** #17780].

[9] S. A. Chobanyan, *The structure of the set of sums of a conditionally convergent series in a Banach space*, Dokl. Akad. Nauk. SSSR **278** (1984), no. 3, 556–559 (Russian); [MR 86h:46020] English transl. in Soviet. Math. Dokl. **30** (1984).

[10] S. A. Chobanyan, *The structure of the set of sums of a conditionally convergent series in a Banach space*, Mat. Sb. **128(170)** (1985), no. 1, 50–65 (Russian); [MR 87c:46014] English transl. in Math. USSR Sb.

[11] W. J. Davis, V. D. Milman, and N. Tomczak-Jaegermann, *The distance between certain n-dimensional Banach spaces*, Israel J. Math. **39** (1981), no. 1-2, 1–15 [MR 82h:46024].

[12] M. M. Day, *Normed Linear Spaces*, 3rd edition, Springer-Verlag, New York-Heidelberg, 1973 [MR **49** #9588].

[13] J. Diestel, *Geometry of Banach Spaces. Selected Topics*, Lecture Notes in Math., Vol. 485, Springer-Verlag, Berlin-New York, 1980 [MR **57** #1079].

[14] J. Diestel, *Sequences and Series in Banach Spaces*, Springer-Verlag, New York-Berlin, 1984 [MR 85i:46020].

[15] V. Dobrić, *A remark on Dvoretzky-Rogers theorem*, Glas. Mat. Ser. III **19** (1984), no. 1, 139–142 [MR 86e:46012].

[16] V. Drobot, *Rearrangements of series of functions*, Trans. Amer. Math. Soc. **142** (1969), 239–248 [MR **39** #7328].

[17] A. Dvoretzky, *Some results on convex bodies and Banach spaces*, in: Proc. Internat. Sympos. Linear Spaces (Jerusalem, 1960), Jerusalem Academic Press, Jerusalem; Pergamon, Oxford, 1961, pp. 123–160 [MR **25** #2518].

[18] A. Dvoretzky and C. Hanani, *Sur les changements des signes des termes d'une série à termes complexes*, C. R. Acad. Sci. Paris **255** (1947), 516–518 [MR **9**, 139g].

[19] A. Dvoretzky and C. A. Rogers, *Absolute and unconditional convergence in normed linear spaces*, Proc. Nat. Acad. Sci. USA **36** (1950), 192–197 [MR **11**, 525a].

[20] P. Enflo, *A counterexample to the approximation problem*, Acta. Math. **130** (1973), 309–317 [MR **53** #6288].

[21] T. Figiel, *A short proof of Dvoretzky's theorem on almost spherical sections of convex bodies*, Compositio Math. **33** (1976), no. 3, 297–301 [MR **58** #7030].

[22] T. Figiel, J. Lindenstrauss, and V. D. Milman, *The dimension of almost spherical sections of convex bodies*, Acta. Math. **139** (1977), no. 1-2, 53–94 [MR **56** #3618].

[23] V. P. Fonf, *Conditionally convergent series in a uniformly smooth Banach space*, Mat. Zametki **11** (1972), no. 2, 209-214 (Russian); [MR **22** #10884] English transl. in Math. Notes **11** (1972).

[24] I. Gel'fand, *Abstrakte Funktionen und lineare Operatoren*, Mat. Sb. **4(46)** (1938), no. 2, 235–286.

[25] D. P. Giesy, *B-convexity and reflexivity*, Israel J. Math. **15** (1973), 430–436 [MR **48** #4698].

[26] D. P. Giesy and R. C. James, *Uniformly non-l^1 and B-convex Banach spaces*, Studia Math. **48** (1973), 61–69 [MR **48** #11994].

[27] V. S. Grinberg and S. V. Sevast'yanov, *Value of the Steinitz constant*, Funkstional. Anal. i Prilozhen. **14** (1980), no. 2, 56–57 (Russian); [MR 81h:52008] English transl. in Functional Anal. Appl. **14** (1980), no. 2.

[28] V. I. Gurarii, M. I. Kadets, and V. I. Matsaev, *Distances between finite-dimensional analogs of the L_p-spaces*, Mat. Sb. **70(112)** (1966), no. 4, 481–489 (Russian); [MR **33** #4649] English transl. in Amer. Math. Soc. Transl. (2) **76** (1968).

[29] R. C. James, *A nonreflexive Banach space that is uniformly nonoctahedral*, Israel J. Math. **18** (1974), 145–155 [MR **50** #8012].

[30] M. I. Kadets, *On a property of broken lines in n-dimensional space*, Uspekhi Mat. Nauk **8** (1953), no. 1, 139–143 (Russian); [MR **14**, 866d].

[31] M. I. Kadets, *On conditionally convergent series in the space L_p*, Uspekhi Mat. Nauk **9** (1954), no. 1, 107–110 (Russian); [MR **15**, 802a].

[32] M. I. Kadets, *Unconditionally convergent series in a uniformly convex space*, Uspekhi Mat. Nauk **11** (1956), no. 5, 185–190 (Russian); [MR **18**, 733e].

[33] M. I. Kadets, *Geometry of normed spaces*, in: Itogi Nauki i Tekhniki. Mat. Analiz **13**, VINITI, Moscow, 1975, pp. 99–127 (Russian); [MR **58** #30064]; English transl. in J. Soviet Math.

[34] M. I. Kadets and K. Wozniakowski, *On series whose permutations have only two sums*, Bull. Polish Acad. Sci. Math. **37** (1989), no. 1-6, 15–21 [MR 92d:46028].

[35] V. M. Kadets, *B-convexity and the Steinitz lemma*, Izv. Severo-Kavkaz. Nauchn. Tsentra Vyssh. Shkoly Estestv. Nauki (1984), no. 4, 27–29 (Russian); [MR 87a:46032].

[36] V. M. Kadets, *A problem of S. Banach (problem* 106 *from the "Scottish Book")*, Funkstional. Anal. i Prilozhen. **20** (1986), no. 4, 74–75 (Russian); [MR 88f:46028] English transl. in Functional Anal. Appl. **20** (1986), no. 4.

[37] V. M. Kadets, *The Steinitz theorem and B-convexity*, Izv. Vyssh. Uchebn. Zaved. **12** (1986), 32–36 (Russian); [MR 88e:46013]; English transl. in Soviet Math. (Izv. VUZ) (1986).

[38] V. M. Kadets, *A domain of sums of weakly convergent series*, Funkstional. Anal. i Prilozhen. **23** (1989), no. 2, 60–62 (Russian); [MR 90j:40004] English transl. in Functional Anal. Appl. **23** (1989), no. 2.

[39] V. M. Kadets, *Rearrangement of function series and different forms of convergence*, Dokl. Akad. Nauk. SSSR **310** (1990), no. 1, 17–20 (Russian); [MR 91a:40002] English transl. in Soviet. Math. Dokl..

[40] V. M. Kadets, *How many points can the range of sums of a series in a Banach space have?*, Teor. Funktsii Funktsional. Anal. i Prilozhen. **54** (1990), 54–57 (Russian); [MR 92c:46012].

[41] V. M. Kadets, *Weak and strong domains of sums of a series in a Banach space*, Mat. Zametki **48** (1990), no. 2, 36–44 (Russian); [MR 91j:46016] English transl. in Math. Notes **48** (1990), no. 1–2.

[42] V. M. Kadets, *On a problem of the existence of a convergent rearrangement*, Izv. Vysh. Uchebn. Zaved. (1992), no. 3, 7–9 (Russian); [MR 94e:40013].

[43] V. M. Kadets and M. I. Kadets, *Rearrangements of Series in Banach Spaces*, Tartu Gos. Univ., Tartu, 1988 [MR 90b:46038]; English transl. Translations of Math. Monographs, **86**, Amer. Math. Society, Providence, R. I., 1991 [MR 92d:46029].

[44] J. P. Kahane, *Some Random Series of Functions*, Heath, Lexington, MA, 1973 [MR **40** #8095].

[45] A. N. Kolmogorov and S. V. Fomin, *Elements of the Theory of Functions and Functional Analysis*, 2nd ed., "Nauka", Moscow, 1988 [MR **38** #2559]; English transl. of 1st ed., vols. I, II, Graylock Press, Albany, New York, 1957, 1961.

[46] P. A. Kornilov, *Rearrangements of conditionally convergent functional series*, Mat. Sb. **113(155)** (1980), no. 4, 598–616 (Russian); [MR 82b:40004] English transl. in Math. USSR Sb. **41** (1982).

[47] P. A. Kornilov, *On the linearity of the set of sums of a functional series*, Uspekhi Mat. Nauk **37** (1982), no. 2, 205–206 (Russian); [MR 83g:40015].

[48] P. A. Kornilov, *The set of sums of a conditionally converging function series*, Mat. Sb. **137(179)** (1988), no. 1, 114–127 (Russian); [MR 89k:40002] English transl. in Math. USSR Sb..

[49] S. Kwapień, *Isomorphic characterizations of inner product spaces by orthogonal series with vector valued coefficients*, Studia Math. **44** (1972), 583–595 [MR **49** #5789].

[50] P. Lévy, *Sur les séries semi-convergentes*, Nouv. Ann. of Math. **5** (1905), 506–511.

[51] J. Lindenstrauss, *On the modulus of smoothness and divergent series in Banach spaces*, Michigan Math. J. **10** (1963), 241–252 [MR **29** #6316].

[52] J. Lindenstrauss and A. Pełczyński, *Absolutely summing operators in L_p-spaces and their applications*, Studia Math. **29** (1968), 275–326 [MR **37** #6743].

[53] J. Lindenstrauss and L. Tzafriri, *Classical Banach Spaces* I, Springer-Verlag, Berlin, 1977 [MR **58** #17766].

[54] J. Lindenstrauss and L. Tzafriri, *Classical Banach Spaces* II, Springer-Verlag, Berlin, 1977 [MR 81c:46001].

[55] L. A. Lyusternik and V. I. Sobolev, *Elements of Functional Analysis*, 2nd revised edition, "Nauka", Moscow, 1965 [MR **35** #698]; English transl. Hindustan Publ. Co., Delhi; Halstead Press, New York, 1974 [MR **50** #2852].

[56] R. P. Maleev, *Conditionally convergent series in some Banach lattices*, C. R. Acad. Bulgare Sci. **32** (1979), no. 8, 1015–1018 (Russian); [MR 81g:46029].

[57] B. Maurey and G. Pisier, *Séries de variables aléatoires indeépendentes et propriétés géométriques des espaces de Banach*, Studia Math. **58** (1976), no. 1, 45–90 [MR **56** #1388].

[58] V. D. Mil'man, *A new proof of A. Dvoretzky's theorem on cross-sections of convex bodies*, Funkstional. Anal. i Prilozhen. **5** (1971), no. 4, 28–37 (Russian); [MR **45** #2451] English transl. in Functional Anal. Appl. **4** (1971), no. 4.

[59] V. D. Milman and M. Sharir, *A new proof of the Maurey-Pisier theorem*, Israel J. Math. **33** (1979), no. 1, 73–87 [MR 83b:46022].

[60] E. M. Nikishin, *The set of sums of a function series*, Mat. Zametki **7** (1970), no. 1, 403–410 (Russian); [MR **42** #6557] English transl. in Math. Notes **7** (1970).

[61] E. M. Nikishin, *Rearrangements of function series*, Mat. Sb. **85(127)** (1971), no. 2, 272–286 (Russian); [MR **44** #1966] English transl. in Math. USSR Sb. **14** (1971).

[62] E. M. Nikishin, *Rearrangements of series in L_p*, Mat. Zametki **14** (1973), no. 1, 31–38 (Russian); [MR **48** # 9172] English transl. in Math. Notes **14** (1973).

[63] W. Orlicz, *Beiträge zu Theorie der Orthogonalentwicklungen*, II, Studia Math. (1929), no. 1, 241–255.

[64] W. Orlicz, *Über unbedingte Konvergenz in Funktionräumen*, Studia Math. (1930), no. 1, 83–85.

[65] M. I. Ostrovskii, *Domains of sums of conditionally covergents series in Banach spaces*, Teor. Funktsii Funktsional. Anal. i Prilozhen. **46** (1986), 77–85 (Russian); [MR 88i:46024].

[66] D. V. Pecherskii, *Rearrangements of the terms in function series*, Dokl. Akad. Nauk. SSSR **209** (1973), no. 6, 1285–1287 (Russian); [MR **47** #9120] English transl. in Soviet. Math. Dokl. **14** (1973).

[67] D. V. Pecherskii, *A theorem on projections of permuted series with terms belonging to L_p*, Izv. Akad. Nauk SSSR Ser. Mat. **41** (1977), no. 1, 203–214 (Russian); [MR **55** #10901] English transl. in Math. USSR Izv. **11** (1977).

[68] D. V. Pecherskii, *Rearrangements of series in Banach spaces and arrangements of signs*, Mat. Sb. **135(177)** (1988), no. 1, 24–35 (Russian); [MR 89f:46023] English transl. in Math. USSR Sb..

[69] A. Pietsch, *Nuclear Locally Convex Spaces*, Springer-Verlag, New York-Heidelberg, 1972 [MR **50** #2853].

[70] A. Pietsch, *Operator Ideals*, North-Holland, Amsterdam-New York, 1980 [MR 81j:47001].

[71] G. Pisier, *Sur les espaces de Banach qui ne contiennent pas uniformément de l_n^1*, C. R. Acad. Sci. Paris **277** (1973), A991–A994 [MR **48** #11998].

[72] G. Pisier, *Martinagales with values in uniformly convex spaces*, Israel J. Math. **20** (1975), no. 3-4, 326–350 [MR **52** #14940].

[73] G. Pisier, *Quotients of Banach spaces of cotype q*, Proc. Amer. Math. Soc. **85** (1982), no. 1, 32–36 [MR 84d:46019].

[74] S. A. Rakov, *Classes of Banach spaces that are connected with unconditional convergence*, Teor. Funktsii Funktsional. Anal. i Prilozhen. **25** (1976), 106–112 (Russian); [MR **55** #3763].

[75] S. A. Rakov, *Banach spaces in which Orlicz's theorem is not valid*, Mat. Zametki **14** (1973), no. 1, 101–106 (Russian); [MR **48** #9373] English transl. in Math. Notes **14** (1973).

[76] A. P. Robertson and W. Robertson, *Topological Vector Spaces*, 2nd edition, Cambridge Univ. Press, London-New York, 1973 [MR **50** #2854].

[77] R. T. Rockafellar, *Convex Analysis*, Princeton Univ. Press, Princeton, N. J., 1970 [MR **43** #445].

[78] W. Rudin, *Functional Analysis*, Second edition, McGraw-Hill, Inc., New York, 1991 [MR 92k:46001].

[79] H. H. Schaefer, *Topological Vector Spaces*, 3rd printing, corrected, Springer-Verlag, New York-Berlin, 1971 [MR **49** #7722].

[80] L. Schwartz, *Geometry and Probability in Banach Spaces*, Lecture Notes in Math., Vol. 852, Springer-Verlag, New York-Berlin, 1981 [MR 83j:46003].

[81] , *The Scottish Book*, Birkhäuser, Boston, MA, 1981 [MR 84m:00015].

[82] S. V. Sevast'yanov, *Approximate solution of some problems in scheduling theory*, in: Methods of Discrete Analysis in the Design of Control Systems, Novosibirsk, 1978, pp. 66–75 (Russian); [MR 82m:90093].

[83] I. Singer, *Bases in Banach Spaces* I, Springer-Verlag, New York-Berlin, 1970 [MR **45** #7451].

[84] I. Singer, *Bases in Banach Spaces* II, Edit. Acad. R. S. Romania, Bucharest; Springer-Verlag, Berlin-New York, 1981 [MR **45** #7451].

[85] D. O. Shklyarskii, *Conditionally convergent series of vectors*, Uspekhi Mat. Nauk (1944), no. 10, 51–59 (Russian); [MR **7**, 12].

[86] C. A. Stegall, *A proof of the principle of local reflexivity*, Proc. Amer. Math. Soc. **78** (1980), no. 1, 154–156 [MR 81e:46012].

[87] E. Steinitz, *Bedingt konvergente Reihen und konvexe Systeme*, J. Reine Angew. Math. **143** (1913), 128–175; **144** (1914), 1–49; **146** (1916), 68-111.

[88] S. J. Szarek, *On the best constants in the Khinchin inequality*, Studia Math. **58** (1976), no. 2, 197–208 [MR **55** #3672].

[89] M. Talagrand, *On the best constants in Khinchin's inequality*, Studia Math. **58** (1976), no. 2, 195–208 [MR **55** #3672].

[90] S. Trojanski, *Conditionally convergent series in certain F-spaces*, Teor. Funktsii Funktsional. Anal. i Prilozhen. **5** (1967), 102–107 (Russian); [MR **37** #3245].

[91] S. Ulam, *A Collection of Mathematical Problems*, Interscience Publ., New York-London, 1960 [MR **22** #10884].

[92] N. N. Vakhaniya, V. I. Tarieladze, and S. A. Chobanyan, *Probability Distributions in Banach Spaces*, "Nauka", Moscow, 1985 (Russian); English transl., 2nd edition, Springer-Verlag, New York-Berlin, 1989 [MR 86j:60014].

[93] Y. Yamasaki, *A simple proof of Kwapień's theorem*, Publ. Res. Inst. Math. Sci. **20** (1984), no. 6, 1247–1251 [MR 86j:46025].

[94] R. M. G. Young, *On the best possible constants in the Khintchine inequality*, J. London Math. Soc. (2) **14** (1976), no. 3, 496–504 [MR **55** #11,008].

Added in proof

[95] S. A. Chobanyan and G. J. Georgobiani, *A problem on rearrangement of summands in normed spaces and Rademacher sums*, in: Probability Theory on Vector Spaces, IV (Lancut, 1987), Lecture Notes in Math., Vol 1391, Springer-Verlag, Berlin-New York, 1989, pp. 33–46; [MR 90m:46018].

[96] E. D. Gluskin, *The diameter of the Minkowski compactum is roughly equal to n*, Funktional. Anal. i Prilozhen. **15** (1981), no. 1, 72–73 (Russian); [MR 83d:46026] Enlish Transl. in Functional Anal. Appl. **15** (1981), no. 1.

[97] H. Hadwiger, *Über das Umordnungsproblem im Hilbertschen Raum*, Math. Zeitschrift **46** (1940), 70–79 [MR **1**, 242].

[98] A. S. Kronrod, *On permutation of terms of numerical series*, Mat. Sb. **18(60)** (1946), 237–280 (Russian); [MR **7**, 432].

[99] V. D. Milman and G. Schechtman, *Asymptotic Theory of Finite-Dimensional Normed Spaces*, Lecture Notes in Math., Vol 1466, Springer-Verlag, Berlin-New York, 1986 [MR 87m:46038].

[100] Sz. Gy. Révész, *Rearrangement of Fourier series*, J. Approx. Theory **60** (1990), no. 1, 101–121 [MR 90m:42042].

[101] P. Wojtaszczyk, *Banach Spaces for Analysts*, Cambridge University Press, Cambridge, 1986 [MR 93d:46001].

INDEX

Titles previously published in the series

OPERATOR THEORY: ADVANCES AND APPLICATIONS

BIRKHÄUSER VERLAG

Edited by
I. Gohberg,
School of Mathematical Sciences, Tel-Aviv University, Ramat Aviv, Israel

This series is devoted to the publication of current research in operator theory, with particular emphasis on applications to classical analysis and the theory of integral equations, as well as to numerical analysis, mathematical physics and mathematical methods in electrical engineering.

45. **J.A. Ball, I. Gohberg, L. Rodman:** Interpolation of Rational Matrix Functions, 1990, (3-7643-2476-7)
46. **P. Exner, H. Neidhardt** (Eds.): Order, Disorder and Chaos in Quantum Systems, 1990, (3-7643-2492-9)
47. **I. Gohberg** (Ed.): Extension and Interpolation of Linear Operators and Matrix Functions, 1990, (3-7643-2530-5)
48. **L. de Branges, I. Gohberg, J. Rovnyak** (Eds.): Topics in Operator Theory. Ernst D. Hellinger Memorial Volume, 1990, (3-7643-2532-1)
49. **I. Gohberg, S. Goldberg, M.A. Kaashoek:** Classes of Linear Operators, Volume I, 1990, (3-7643-2531-3)
50. **H. Bart, I. Gohberg, M.A. Kaashoek** (Eds.): Topics in Matrix and Operator Theory, 1991, (3-7643-2570-4)
51. **W. Greenberg, J. Polewczak** (Eds.): Modern Mathematical Methods in Transport Theory, 1991, (3-7643-2571-2)
52. **S. Prössdorf, B. Silbermann:** Numerical Analysis for Integral and Related Operator Equations, 1991, (3-7643-2620-4)
53. **I. Gohberg, N. Krupnik:** One-Dimensional Linear Singular Integral Equations, Volume I, Introduction, 1992, (3-7643-2584-4)
54. **I. Gohberg, N. Krupnik:** One-Dimensional Linear Singular Integral Equations, Volume II, General Theory and Applications, 1992, (3-7643-2796-0)
55. **R.R. Akhmerov, M.I. Kamenskii, A.S. Potapov, A.E. Rodkina, B.N. Sadovskii:** Measures of Noncompactness and Condensing Operators, 1992, (3-7643-2716-2)
56. **I. Gohberg** (Ed.): Time-Variant Systems and Interpolation, 1992, (3-7643-2738-3)
57. **M. Demuth, B. Gramsch, B.W. Schulze** (Eds.): Operator Calculus and Spectral Theory, 1992, (3-7643-2792-8)

58. **I. Gohberg** (Ed.): Continuous and Discrete Fourier Transforms, Extension Problems and Wiener-Hopf Equations, 1992, (3-7643-2809-6)
59. **T. Ando, I. Gohberg** (Eds.): Operator Theory and Complex Analysis, 1992, (3-7643-2824-X)
60. **P.A. Kuchment:** Floquet Theory for Partial Differential Equations, 1993, (3-7643-2901-7)
61. **A. Gheondea, D. Timotin, F.-H. Vasilescu** (Eds.): Operator Extensions, Interpolation of Functions and Related Topics, 1993, (3-7643-2902-5)
62. **T. Furuta, I. Gohberg, T. Nakazi** (Eds.): Contributions to Operator Theory and its Applications. The Tsuyoshi Ando Anniversary Volume, 1993, (3-7643-2928-9)
63. **I. Gohberg, S. Goldberg, M.A. Kaashoek:** Classes of Linear Operators, Volume 2, 1993, (3-7643-2944-0)
64. **I. Gohberg** (Ed.): New Aspects in Interpolation and Completion Theories, 1993, (3-7643-2948-3)
65. **M.M. Djrbashian:** Harmonic Analysis and Boundary Value Problems in the Complex Domain, 1993, (3-7643-2855-X)
66. **V. Khatskevich, D. Shoiykhet:** Differentiable Operators and Nonlinear Equations, 1993, (3-7643-2929-7)
67. **N.V. Govorov †:** Riemann's Boundary Problem with Infinite Index, 1994, (3-7643-2999-8)
68. **A. Halanay, V. Ionescu:** Time-Varying Discrete Linear Systems Input-Output Operators. Riccati Equations. Disturbance Attenuation, 1994, (3-7643-5012-1)
69. **A. Ashyralyev, P.E. Sobolevskii:** Well-Posedness of Parabolic Difference Equations, 1994, (3-7643-5024-5)
70. **M. Demuth, P. Exner, G. Neidhardt, V. Zagrebnov** (Eds): Mathematical Results in Quantum Mechanics. International Conference in Blossin (Germany), May 17–21, 1993, 1994, (3-7643-5025-3)
71. **E.L. Basor, I. Gohberg** (Eds): Toeplitz Operators and Related Topics. The Harold Widom Anniversary Volume. Workshop on Toeplitz and Wiener-Hopf Operators, Santa Cruz, California, September 20–22, 1992, 1994 (3-7643-5068-7)
72. **I. Gohberg, L.A. Sakhnovich** (Eds): Matrix and Operator Valued Functions. The Vladimir Petrovich Potapov Memorial Volume, (3-7643-5091-1)
73. **A. Feintuch, I. Gohberg** (Eds): Nonselfadjoint Operators and Related Topics. Workshop on Operator Theory and Its Applications, Beersheva, February 24–28, 1994, (3-7643-5097-0)
74. **R. Hagen, S. Roch, B. Silbermann:** Spectral Theory of Approximation Methods for Convolution Equations, 1994, (3-7643-5112-8)
75. **C.B. Huijsmans, M.A. Kaashoek, B. de Pagter**: Operator Theory in Function Spaces and Banach Lattices. The A.C. Zaanen Anniversary Volume, 1994 (ISBN 3-7643-5146-2)
76. **A.M. Krasnosellskii**: Asymptotics of Nonlinearities and Operator Equations, 1995, (ISBN 3-7643-5175-6)

77. **J. Lindenstrauss, V.D. Milman** (Eds): Geometric Aspects of Functional Analysis Israel Seminar GAFA 1992–94, 1995, (ISBN 3-7643-5207-8)
78. **M. Demuth, B.-W. Schulze** (Eds): Partial Differential Operators and Mathematical Physics: International Conference in Holzhau (Germany), July 3–9, 1994, 1995, (ISBN 3-7643-5208-6)
79. **I. Gohberg, M.A. Kaashoek, F. van Schagen**: Partially Specified Matrices and Operators: Classification, Completion, Applications, 1995, (ISBN 3-7643-5259-0)
80. **I. Gohberg, H. Langer** (Eds): Operator Theory and Boundary Eigenvalue Problems. International Workshop in Vienna, July 27–30, 1993, 1995, (ISBN 3-7643-5275-2)
81. **H. Upmeier**: Toeplitz Operators and Index Theory in Several Complex Variables, 1996, (ISBN 3-7643-5282-5)
82. **T. Constantinescu**: Schur Parameters, Factorization and Dilation Problems, 1996, (ISBN 3-7643-5285-X)
83. **A.B. Antonevich**: Linear Functional Equations. Operator Approach, 1995, (ISBN 3-7643-2931-9)
84. **L.A. Sakhnovich**: Integral Equations with Difference Kernels on Finite Intervals, 1996, (ISBN 3-7643-5267-1)

85/ **Y.M. Berezansky, G.F. Us, Z.G. Sheftel**: Functional Analysis, Vol. I + Vol. II, 1996,
86. Vol. I (ISBN 3-7643-5344-9), Vol. II (3-7643-5345-7)

87. **I. Gohberg, P. Lancaster, P.N. Shivakumar** (Eds): Recent Developments in Operator Theory and Its Applications. International Conference in Winnipeg, October 2–6, 1994, 1996, (ISBN 3-7643-5414-5)
88. **J. van Neerven** (Ed.): The Asymptotic Behaviour of Semigroups of Linear Operators, 1996, (ISBN 3-7643-5455-0)
89. **Y. Egorov, V. Kondratiev**: On Spectral Theory of Elliptic Operators, 1996, (ISBN 3-7643-5390-2)
90. **A. Böttcher, I. Gohberg** (Eds): Singular Integral Operators and Related Topics. Joint German-Israeli Workshop, Tel Aviv, March 1–10, 1995, 1996, (ISBN 3-7643-5466-6)
91. **A.L. Skubachevskii**: Elliptic Functional Differential Equations and Applications, 1997, (ISBN 3-7643-5404-6)
92. **A.Ya. Shklyar**: Complete Second Order Linear Differential Equations in Hilbert Spaces, 1997, (ISBN 3-7643-5377-5)
93. **Y. Egorov, B.-W. Schulze**: Pseudo-Differential Operators, Singularities, Applications, 1997, (ISBN 3-7643-5484-4)

GPSR Compliance

The European Union's (EU) General Product Safety Regulation (GPSR) is a set of rules that requires consumer products to be safe and our obligations to ensure this.

If you have any concerns about our products, you can contact us on ProductSafety@springernature.com

In case Publisher is established outside the EU, the EU authorized representative is:

Springer Nature Customer Service Center GmbH
Europaplatz 3
69115 Heidelberg, Germany

Batch number: 10607214

Printed by Printforce, the Netherlands